D1256313

AN INTRODUCTION TO MULTIVARIATE TECHNIQUES
FOR SOCIAL AND BEHAVIOURAL SCIENCES

An Introduction to Multivariate Techniques for Social and Behavioural Sciences

Spencer Bennett
Department of Psychology
University of Bradford

and

David Bowers
Department of Economics
University of Bradford

A Halsted Press Book

JOHN WILEY & SONS
New York

First published in the United Kingdom 1976 by
The Macmillan Press Ltd

Published in the U.S.A. by
Halsted Press, a Division of
John Wiley & Sons, Inc., New York

Printed in Great Britain

Library of Congress Cataloging in Publication Data

Bennett, Spencer.
An introduction to multivariate techniques for social and behavioural sciences.

"A Halsted Press book."
Bibliography: p.
Includes index.
1. Multivariate analysis. 2. Social sciences – Statistical methods. I. Bowers, David, joint author. II. Title.
HA33.B623 519.5′3 74-20108
ISBN 0 470-09280-7

Contents

Preface

Mathematical sophistication involved

Almost without exception, the existing texts covering multivariate statistical techniques are couched in terms which are too complex for the mathematically unsophisticated reader. Despite the undoubted importance, general applicability and power of the different multivariate methods, they are not as widely used in research or taught in undergraduate and postgraduate courses as they should be. It seemed to the authors that one major reason for this is the level of mathematical sophistication required for understanding the existing texts. In the experience of the authors, the chief stumbling block in this respect is the use of matrix algebra in the derivation of the proofs underlying such technique. Not only do mathematically unsophisticated readers have difficulty dealing with matrix algebra, but they have difficulty separating out the basic ideas, assumptions, and the interpretation of results, which are essential for an adequate understanding and use of the methods, from those parts of the exposition which have to do with derivations of expressions used in the techniques, which are not strictly essential.

The advantage of using matrix algebra is that, in general, the exposition of some points can be made shorter and the proof of many relationships facilitated. However, it was the authors' opinion (with considerable experience teaching in this area) that the overwhelming disadvantage of using matrix algebra was the barrier it presented to many (if not most) readers who are unable or unwilling to leap over this first hurdle and for whom, therefore, the field of multivariate analysis must remain unfamiliar. In the experience of the authors the number of social and behavioural scientists (both students and practitioners) who are completely at ease with matrix algebra represents a minority. Even to those with some knowledge of matrix algebra, the use of matrix terminology is often confusing and disguises to some extent what is going on. Furthermore, the authors are not concerned here to present proofs of equations required in the book. On the whole, therefore, the authors preferred understanding at some small cost in length of exposition to conciseness and incomprehensibility. It must be added of course that there are several books of a more advanced character which make free use of matrix algebra, to which our readers might turn with some success after having mastered the essential elements of the various techniques described here.

There exist texts dealing with elementary statistical techniques that can be understood by unsophisticated readers which concentrate on the application of methods rather than on detailed derivations and proofs. Good texts of this kind recognise that it is essential for the reader to (*a*) understand the assumptions he is making when carrying out a statistical test, (*b*) be able to choose a technique that is appropriate for his data or hypotheses, (*c*) be able to execute the procedure correctly, (*d*) be able to interpret the results correctly, and (*e*) recognise the existence of controversies or differences of opinion in any of the above areas. Moreover, such texts equally realise that many of the essential ideas can be presented without confusing the reader with detailed derivations, often expressed in complex mathematical terms. A criticism of many such texts is that they adopt a 'cookbook' approach. There are books which fit such a description, but the better elementary texts are unfairly described thus, because great efforts are made to put over the essential ideas and give the reader a clear, if intuitive, grasp of them.

Many such books exist covering elementary statistics, but almost none exist in the field of multivariate statistics. The aim of the present text is to fill this gap by cutting down the use of matrix algebra, derivations and proofs, whilst giving the reader a clear intuitive grasp of the ideas essential to the understanding and use of the methods.

One justification for such an approach is that one of the books* that does provide an elementary introduction to certain multivariate methods has received a very good response from students, researchers and lecturers in the behavioural sciences who recognise the value of teaching multivariate methods but either (*a*) did not feel sufficiently well-equipped mathematically to undertake such a task, or (*b*) did not feel that their students were able to grasp the essential ideas with the use of current texts.

The aims of the present text are similar in some respects to that of Child, in that a fairly elementary introduction is presented, as far as possible without using matrix algebra, or indeed anything more complex than manipulation of simple algebraic expressions. However, the scope is considerably wider, covering techniques other than factor analysis to which Child confines himself.

Use of numerical examples

In the present text, extensive use has been made of numerical examples, often carried out in complete detail, so that a reader could, if he wished, perform many of the analyses by hand using a desk calculator. In practice, with actual research data, the reader would not wish to do this, since the computational labour involved is formidable, and, in any case, computer programmes for all the tech-

*D. Child, *Essentials of Factor Analysis* (New York: Holt, Rinehart & Winston, 1970).

niques are available. Since presenting detailed numerical examples takes up extra space which could be devoted to the extension of the book elsewhere, a brief justification of this approach is in order.

(1) In the experience of the authors, numerical examples provide one of the best ways of conveying to the mathematically unsophisticated reader what is emerging at each stage of an analysis. The authors, both of whom teach the application of statistical methods to social and behavioural scientists, have found this to be the case. Numerical examples appear to concretise the formal operations involved, and many people find it easier when information is presented in this form. Students are always impatient to be given a worked example and it usually seems to be at this stage that the 'light dawns'.
(2) Readers may often wish to carry through the analyses for themselves in order to make sure that they *do* understand the steps involved, or a researcher may wish to carry out a pilot study on a small scale for which computation with a desk calculator is feasible. Digging out the computational details from a mass of theoretical algebraic expressions often proves a daunting task for them.

Notes on organisation of the text

With the advent of programmed instruction techniques, where the areas of difficulty experienced by the learner can be isolated by feedback from him, it has become apparent that what appears to the specialist to be the logical order of development in subjects such as mathematics and statistics (and, indeed, many other subjects) is very often not that which is grasped most readily by the learner. A possible criticism of the present text by a specialist might be that the most 'logical' development of the chapters devoted to factor analysis have not been followed. For instance, it may be said that the most logical development is from principal components to principal factor analysis and then to more general methods of factor analysis. Actually, we have begun with the centroid method, which, in our opinion, is much easier to grasp than principal components and principal factor analysis, and hence appears the least confusing way to begin. It may be argued that the centroid method is now only of historic interest, and that is indeed the case, but the computational steps involved are straightforward, so the reader can easily work through a simple example for himself and thus more quickly gain an understanding of the basic aims of factor-analytic methods.

In a book such as this, it is difficult to maintain a constant level of simplicity of exposition in the presentation of each method, and the assumptions concerning the desired level of knowledge of the reader for each chapter probably vary somewhat. Overcoming this problem within the text would make the book longer than the authors wish, and although every effort has been made to achieve uniformity, there may be some chapters which a certain proportion of readers will find less simple than others (as is indeed the case with any text). Nevertheless, although this may mean that, for certain chapters, understanding of some details may be lost, much of the essential information for understanding the techniques should be within the scope of all readers with some basic knowledge of elementary statistics.

The scope of the text is broader than most other texts on multivariate statistics. A wide range of potentially useful techniques have been included, although we have reluctantly had to exclude some topics in the interests of keeping the book within reasonable bounds. Some areas, such as multiple regression analysis, multivariate analysis of variance and canonical correlation have been left out for this reason. However, good accounts of these techniques can be found elsewhere, for example in Kerlinger and in Baggaley.*

*F. N. Kerlinger and E. J. Pedhazur, *Multiple Regression in Behavioural Research* (New York: Holt, Rinehart & Winston, 1973) and Andrew R. Baggaley, *Intermediate Correlational Methods* (New York: Wiley, 1964).

1 Introduction

1.1 Preliminary remarks

In recent years a number of powerful techniques (in addition to the more traditional methods of regression, correlation, and analysis of variance, for example) have been used by researchers in the social and behavioural sciences. These techniques, known collectively as multivariate analysis, as versatile and wide-ranging in application as they are, deserve to be known to a wider audience than hitherto, and it is the purpose of this book to introduce a selection of the most generally useful of them in as simple a manner as possible.

Unfortunately, most of the existing texts are couched in the language of advanced mathematics (for example, matrix algebra) much beyond the ability of, or long forgotten by, many people working in the social and behavioural science areas, both academic and professional. The authors believe that this book will appeal to those people who want a 'gentle' introduction to the methods of multivariate analysis. It demands no more of its readers than a knowledge of some relatively elementary algebraic techniques (e.g. the manipulation of simple equations and some familiarity with summation notation) and some knowledge of basic statistics (for example, the concepts of correlation, variance, normal distribution, simple probability and the meaning of statistical significance) which is to be found in a great number of introductory texts.* The authors have tried to avoid derivations or proofs (leaving these to more advanced texts) and the emphasis throughout has been to provide an intuitive grasp of the broad aims of each technique, together with a sound understanding of how each may be applied. The assumptions underlying the use of the techniques and their limitations are also discussed.

Because of this approach, even the reader without the above limited knowledge should have little difficulty in gaining an appreciation of the essentials of each of the methods discussed.

It is hoped that a mathematically unsophisticated researcher who wishes to conduct an investigation and answer some questions will, with the aid of this book, know in what form his data should be, which multivariate analytic technique is the most appropriate, how to apply it and how to interpret any results he obtains.

*For example, Hoel [1971] or Yamane [1964].

It is unlikely that the researcher will in practice be able to perform the necessary computations himself, since most real problems will be too unmanageable to be solved manually and the necessary work will undoubtedly be carried out on a computer (unless it is thought that a limited-range pilot study with manual computation would profitably improve understanding of the application of any of the techniques to a particular problem). Nevertheless, the authors believe that the efficient and successful use of computers in the solution of such problems, and meaningful interpretation of the output, makes it desirable for the reader to have some appreciation of the steps involved in the solution procedures and therefore the book gives these steps in detail.

A large part of this book is concerned with a group of multivariate techniques known generically as factor analysis. One of the differences between well-established physical sciences and the behavioural sciences is that, in the former, the area of knowledge has been so extensively studied that the relationships between the variables manipulated or observed are often well known and many of the important ones have already been identified. The social and behavioural scientist is not in such a happy position. Very often he neither knows which are the important variables to study, nor understands the relationships between them. Furthermore, the variables he can observe are often only loosely related to those which are measures of the concepts that he is really interested in, just as 'height' is only indirectly a measure of 'body size', and 'income' is only indirectly a measure of 'socio-economic status'. He knows that he must make do with several indirect (or impure) measures and combine them in various complex ways in order to obtain a more refined measure of the concepts which he deems important. Thus the behavioural scientist moves in an intellectual world inhabited by a confused morass of more or less interlocking or interdependent variables and vague concepts that resist direct measurement. In this respect, factor analysis is a powerful tool, and can help him to bring some order into this chaos. By helping him to replace a large set of observable variables by fewer unobserved constructs (taking advantage of the fact that so many behavioural variables are interdependent), factor analysis helps him to provide a parsimonious description of his domain of study. Because of its importance, factor analysis is the sole concern of three chapters (2, 3 and 4) and is used in the analysis of data obtained in special, relatively unfamiliar ways in three other chapters (5, 6 and 7).

Another central concern of some multivariate techniques is to provide the best way of discriminating between several groups (or samples) on the basis of differences in their scores on several variables. The well-known analysis of variance can be used for this purpose in the case where only one variable is used for distinguishing between several groups, but when more than one variable is used for this purpose special multivariate techniques are required. These techniques also deal with the problem of classification, i.e. how to use measures on several variables as a basis for decisions about which group individuals should be allocated to. These techniques are discussed in Chapters 6 and 7.

Most of the techniques to be discussed ideally require variables to be measureable quantitatively and, in some cases, to be at least approximately normally distributed. However, many variables with which social and behavioural scientists are concerned are qualitative (e.g. sex or religion). A selection of useful multivariate techniques applicable to such data is discussed at length in Chapter 7.

In addition to being a relatively simple account of some of the most widely used multivariate techniques, this book brings together a selection of such techniques that are normally found in separate texts, and illustrates the relationships between the various techniques, some of which can be used in combination with one another.

1.2 Correlation

Many of the techniques to be described begin with sets of correlations between variables and therefore some points must be made concerning the assumptions made about the distributions of the variables which must hold if correlations are to be used successfully and with reliability in the results. First of all, the variables should be continuous, and must be linear functions of each other (in this last respect, for example, the relationship in a bivariate regression model should be a straight line). Although multivariate normalcy in the distribution is desirable, perhaps it is more important that there should not be any marked skewness which would lead to unreliable results. Reliability and stability (i.e. insensitivity to small changes in the data) in the results will increase asymptotically as sample size increases and for this reason real efforts should be made to achieve sample sizes as large as possible; for the purposes of factor analysis, sample sizes of several hundred are advisable. Following the usual conventions, r_{ij} represents the correlation between variables i and j. The square of the correlation, r_{ij}^2 is effectively a measure of the *common* variance between variables. The total variance possessed by a variable in its relationship with other variables can in fact be decomposed into three components, common variance (which is shared with other variables), unique variance (that which is unique to that variable and not explained by other variables), and error variance, which is due to error in measurement. This is probably familiar to the reader with a basic background in elementary statistics, but in any case it will be explained further in the next chapter.

1.3 Matrices

Although the book does not require knowledge of matrix algebra, it is necessary to know that a matrix is simply a rectangular array of numbers ordered in some special way. For example, a football league table is a matrix with the rows corresponding to the teams (which might be listed down the left) and the columns, perhaps to games played, points scored, etc. The correlation matrix will be a square array of numbers, each of which will give the correlation between two variables, the variables listed down the left-hand side and across the top of the table. A correlation matrix R, giving the correlations between three variables, X_1, X_2, and X_3, might look like this:

		variables			
		X_1	X_2	X_3	
	X_1	1.0	0.2	0.3	
variables	X_2	0.2	1.0	0.6	*A correlation matrix, R with elements* r_{ij}
	X_3	0.3	0.6	1.0	

The rows of a matrix are usually denoted with the letter i, the columns with j, and an element of the above correlation matrix might be referred to as r_{ij}, meaning the correlation to be found in the ith row and jth column. Thus r_{23} would refer to the element with value 0.6 in the second row and third column, being the correlation between X_2 and X_3. Notice that this correlation matrix is 'symmetric', i.e. the values in the upper N.E. triangle are the same as those in the lower S.W. triangle; this is because the correlation between X_2 and X_3 is the same as that between X_3 and X_2, of course. The elements r_{11}, r_{22}, and r_{33} take the value 1.00 since they are the correlation of each variable with itself, and elements in these positions in the matrix are referred to as being in the 'principal diagonal' of the matrix. In actual fact, we noted earlier that the total variance of a variable is made up of three components, one of which is 'error variance', produced by the fact that the measurement is rarely perfect, so that two sets of measurements on the same variable would hardly ever correlate perfectly. Putting 1's in the principal diagonal therefore assumes that no error of measurement is involved.

1.4 Matrix inverse: the square-root method

No matrix algebra will be required for understanding this text, so the reader need not be frightened by the following section and, in fact, if desired it can be skipped altogether. However, in one of the following chapters, the operation known in matrix algebra as finding the inverse of a matrix is required, and so, for the interested reader, the details of the procedure are described here. If the matrix is denoted by A, say, then the inverse is denoted conventionally as A^{-1}. This is analogous to taking the inverse of an ordinary number although, because of the complexities of matrix algebra, the process is much more difficult (one cannot divide one matrix by another). In practice, any necessary matrix inverse procedure will be carried out for the researcher within the computer program he is using for some particular multivariate technique. However, a simple algebraic scheme for inverting a small matrix, known as the square root method, is presented here, so that the reader, if he wishes, can refer to it when the method is encountered later in the text. The method in the case of a matrix with four rows and columns is as follows:

Suppose we wish to find the inverse of a symmetric matrix A, with elements a_{ij}, viz.

$$A = \begin{bmatrix} a_{11} & a_{12} & a_{13} & a_{14} \\ & a_{22} & a_{23} & a_{24} \\ & & a_{33} & a_{34} \\ & & & a_{44} \end{bmatrix}$$

(The elements in the lower triangle below the principal diagonal will be the same as those in the upper triangle and are not shown here.)

The matrix A is shown below in the upper left-hand quadrant of a set of four matrices.

1	a_{11}	a_{12}	a_{13}	a_{14}	1	0	0	0
2		a_{22}	a_{23}	a_{24}	0	1	0	0
3			a_{33}	a_{34}	0	0	1	0
4				a_{44}	0	0	0	1
1a	s_{11}	s_{12}	s_{13}	s_{14}	s_{15}	0	0	0
2a		s_{22}	s_{23}	s_{24}	s_{25}	s_{26}	0	0
3a			s_{33}	s_{34}	s_{35}	s_{36}	s_{37}	0
4a				s_{44}	s_{45}	s_{46}	s_{47}	s_{48}

The matrix to the right of A is known as an identity matrix (usually denoted by the letter I) and is analogous to the number 1 of ordinary arithmetic. It consists of ones in the principal diagonal and zeros elsewhere. We establish the two bottom matrices using the following relationships:

For row 1a:

$$s_{11} = a_{11}/\sqrt{a_{11}}; \qquad s_{12} = a_{12}/\sqrt{a_{11}}; \qquad s_{13} = a_{13}/\sqrt{a_{11}};$$

$$s_{14} = a_{14}/\sqrt{a_{11}}; \qquad s_{15} = a_{15}/\sqrt{a_{11}}.$$

For row 2a:

$$s_{22} = \sqrt{a_{22} - s_{12}s_{12}}; \qquad s_{23} = (a_{23} - s_{12}s_{13})/s_{22};$$

$$s_{24} = (a_{24} - s_{12}s_{14})/s_{22}; \qquad s_{25} = (0.0 - s_{12}s_{15})/s_{22};$$

$$s_{26} = (1.0 - s_{12}\,0.0)/s_{22}.$$

For row 3a:

$$s_{33} = \sqrt{a_{33} - s_{13}s_{13} - s_{23}s_{23}}$$

$$s_{34} = (a_{34} - s_{13}s_{14} - s_{23}s_{24})/s_{33}$$

$$s_{35} = (0.0 - s_{13}s_{15} - s_{23}s_{25})/s_{33}$$

$$s_{36} = (0.0 - s_{13}\,0.0 - s_{22}s_{26})/s_{33}$$

$$s_{37} = (1.0 - s_{13}\,0.0 - s_{23}\,0.0)/s_{33}$$

For row 4a:

$$s_{44} = \sqrt{a_{44} - s_{14}s_{14} - s_{24}s_{24} - s_{34}s_{34}}$$

$$s_{45} = (0.0 - s_{14}s_{15} - s_{24}s_{25} - s_{34}s_{35})/s_{44}$$

$$s_{46} = (0.0 - s_{14}\,0.0 - s_{24}s_{26} - s_{34}s_{36})/s_{44}$$

$$s_{47} = (0.0 - s_{14}\,0.0 - s_{24}\,0.0 - s_{34}s_{37})/s_{44}$$

$$s_{48} = (1.0 - s_{14}\,0.0 - s_{24}\,0.0 - s_{34}\,0.0)/s_{44}$$

Then the elements of a_{ij}^{-1} the inverse A^{-1} of A, i.e.

$$A^{-1} = \begin{bmatrix} a_{11}^{-1} & a_{12}^{-1} & a_{13}^{-1} & a_{14}^{-1} \\ a_{21}^{-1} & a_{22}^{-1} & a_{23}^{-1} & a_{24}^{-1} \\ a_{31}^{-1} & a_{32}^{-1} & a_{33}^{-1} & a_{34}^{-1} \\ a_{41}^{-1} & a_{42}^{-1} & a_{43}^{-1} & a_{44}^{-1} \end{bmatrix}$$

are given:

$$a_{11}^{-1} = s_{15} + s_{25}^2 + s_{35}^2 + s_{45}^2;$$

$$a_{12}^{-1} = s_{25}s_{26} + s_{35}s_{36} + s_{45}s_{46};$$

$$a_{13}^{-1} = s_{35}s_{37} + s_{45}s_{47}; \quad a_{14}^{-1} = s_{45}s_{48};$$

$$a_{21}^{-1} = s_{26}s_{25} + s_{36}s_{35} + s_{46}s_{45}; \quad a_{22}^{-1} = s_{26}^2 + s_{36}^2 + s_{46}^2;$$

$$a_{23}^{-1} = s_{36}s_{37} + s_{46}s_{47}; \quad a_{24}^{-1} = s_{46}s_{48};$$

$b^{-1}_{31} = s_{37}s_{35} + s_{47}s_{45}$; $b^{-1}_{32} = s_{37}s_{36} + s_{47}s_{46}$;

$b^{-1}_{33} = s^2_{37} + s^2_{47}$; $b_{34} = s_{47}s_{48}$; $b_{41} = s_{48}s_{45}$;

$b_{42} = s_{48}s_{46}$; $b_{43} = s_{48}s_{47}$; $b_{44} = s^2_{48}$.

The inverse of a symmetric matrix is in fact itself a symmetric matrix and thus the lower triangular elements of A^{-1} are equal to the corresponding upper triangular values computed above, e.g. $a^{-1}_{31} = a^{-1}_{13}$. The above method can be generalised quite simply to square matrices of sizes other than four rows and four columns.

Should the researcher require to employ a computer program for finding the inverse of a matrix, he can find details of the procedure in, for example, Overall and Klett [1972].

2 Factor Analysis: the Centroid Method

2.1 Introduction

We begin with a matrix or table of correlations between a set of variables. Since we are employing product-moment correlations as our starting-point, it is important that the assumptions underlying their use are met. To re-iterate briefly: ideally, the distributions of the variables should be continuous and reasonably normal (at least they should not be bimodal or markedly skewed); discontinuous variables will be discussed further in Chapter 7. Regression should be linear, and samples should be large (at least several hundred) to ensure reliability of the resulting correlations.

Other considerations are that (1) variables should not be linear combinations of each other (e.g. a maths mark, an English mark and an aggregate mark as three separate variables): such variables are apt to lead to the emergence of spurious factors in the analysis; and (2) choosing a sample that is too heterogeneous such that the different groups are liable to give different results if factor analysis were to be carried out on them separately. These points should be reviewed again at the end of the chapter, after the reader has acquired a clearer idea of what factor analysis is about, so that their implications can be more firmly grasped.

The object of factor analysis is fairly simple in principle. Suppose we carry out a set of measurements on a sample of people or objects, then a description of any subject in the sample involves stipulating its score on each of the measures we employ. If we employ ten measurements, then ten scores are required to specify a description of the individual. However, since some of the measures may correlate with (i.e. be predictable from) others then this description is probably uneconomical, since it is possible that the individual may be adequately specified by using a smaller set of variables. The main aim may therefore be said to be *parsimony of description.*

As an illustration of a factor analysis, we have chosen an adaptation of a piece of research conducted by Burt and Banks [1941] on the measurement of nine body dimensions. The values of the initial correlations are not the same as in the original study, but the main conclusions drawn are. The reader may notice at the outset that one of the tenets described above has been violated in this study, i.e. one variable (standing height) is a linear combination of two of the others ('sitting height' and 'leg length'). The way this can affect the interpretation of the results of the analysis will be discussed later.

2 400 male volunteers were measured on nine variables and each measure was inter-correlated to produce the matrix shown in Table 2.1 in which reasonably high correlations are shown in italics.

TABLE 2.1 *Matrix of inter-correlations in body-dimension study*

Variable		1	2	3	4	5	6	7	8
Standing height	1								
Sitting height	2	*0.81*							
Arm length	3	0.59	*0.67*						
Leg length	4	0.58	0.29	0.19					
Thigh length	5	0.53	0.25	0.16	*0.67*				
Abdomen girth	6	0.17	0.13	0.08	0.23	0.29			
Hip girth	7	0.33	0.39	0.28	0.17	0.22	*0.70*		
Shoulder girth	8	0.22	0.29	0.21	0.08	0.12	0.52	0.59	
Weight	9	0.40	0.41	0.29	0.28	0.33	*0.77*	*0.83*	*0.62*

Thus, in terms of the measures chosen, a description of the physical dimensions of any individual requires knowledge of his score on nine variables. But, if we look at the high correlations, 1 (standing height) and 2 (sitting height) appear to go together, as do 2 and 3 (arm length); even 1 and 3 are reasonably closely related. In other words, 1, 2 and 3 appear to form a group of closely related measures and we suspect that it may not be necessary to specify all three since they may represent somewhat different aspects of some underlying variable – say 'body length'. Similar inspection would reveal other potential groupings which leads us to suspect that a more economical description of bodily dimensions is possible in terms of a smaller set of underlying variables. For instance, 4 and 5 appear to form another group, and 6, 7, 8 and 9 another. Perhaps the former may be denoted as 'length below the waist' and the latter as 'body mass'.

Casual inspection of this sort is not very satisfactory since it is unsystematic and ambiguous, but at least it gives us an initial intuitive grasp of what factor analysis is all about. Very roughly, the smaller set of 'underlying variables' are the 'factors' we are interested in and the factor analysis is a formal method of specifying how many there are and the extent to which each original variable is in fact a measure of these 'underlying variables'.

After performing a factor analysis (by the centroid method to be described below) we obtain a factor matrix of the form shown in Table 2.2.

TABLE 2.2 *Factor matrix for nine body dimensions (significant loadings in italic)*

Variable	Factor F_1	F_2	F_3	h^2
Standing height	+ *0.76*	+ *0.50*	+ 0.12	0.84
Sitting height	+ *0.69*	+ *0.41*	+ *0.49*	0.90
Arm length	+ *0.54*	+ *0.33*	+ *0.48*	0.63
Leg length	+ *0.54*	+ *0.48*	− *0.43*	0.70
Thigh length	+ *0.56*	+ *0.31*	− *0.48*	0.64
Abdomen girth	+ *0.63*	− *0.49*	− 0.29	0.72
Hip girth	+ *0.74*	− *0.51*	+ 0.07	0.82
Shoulder girth	+ *0.56*	− *0.43*	+ 0.13	0.53
Weight	+ *0.82*	− *0.40*	− 0.02	0.83

The factor analysis has revealed that there appear to be three 'underlying variables' or factors denoted in Table 2.2 by the columns F. Before we try to identify them, it should be noted that the figures in the body of Table 2.2 represent correlation coefficients between the initial variables and the factors – these coefficients are called *factor loadings* and, like ordinary correlation coefficients, can be regarded as significant (in the statistical sense) or non-significant. For example, there is a correlation between standing height and the first factor of 0.76. The next question to be decided is which of these factor loadings *are* statistically significant. Although we will discuss the significance of factor loadings here it will become apparent in a later section on rotation of factors that this question of significance is fraught with difficulty. There are several different criteria which we may use; a simple rule of thumb procedure is to allow that the loading of the variable on the factor is significant if the loading is ± 0.3 or more. This is not based on any rigorous statistical foundation but is a conservative criterion which ensures that only those factors having a reasonably strong association are identified as being significant. Providing the sample is reasonably large this procedure should prove satisfactory; it has the virtue of being quick and simple to use but does have the disadvantage that a significant factor loading may be neglected; however, it is better perhaps to err on the side of caution.

Tests of significance founded more firmly in statistical theory may be found in Burt and Banks [1947] and Harman [1967]. Using the 0.3 criterion of significance, significant factor loadings have been shown in italics in Table 2.2, and in the next section it will be shown how significant loadings may be used in the interpretation of factors.

2.2 Interpretation of factors

It is not usual (or even desirable) to finally interpret the factors until after their rotation (which will be dealt with later) and sometimes not even then. There is a certain amount of controversy over whether unrotated factors should be interpreted at all, but it may be enlightening to briefly examine the possible meaning of the factor loadings at this stage and return to the problem of interpretation in more detail after looking at the question of rotation.

Inspection of the first factor loadings (F_1) shows that they all appear to be significant and positive. This would *seem* to indicate that there is an underlying attribute common to all the variables, perhaps 'general body size'; that is, all the variables *seem* to be measures of the same thing to greater or lesser extents. 'Weight' is apparently the best measure of this general factor since it has the highest factor loading. The word 'seem' has been emphasised above because, firstly, the method of factor analysis employed ensures that the first factor will have this general character, as will be explained later. Secondly, the factor is made up of variables that may bear little relationship (be uncorrelated) with each other; for example, variables 4 (leg length) and 8 (shoulder girth) have a correlation of only 0.08 (i.e. have only 0.6 per cent variance in common) and yet both are highly loaded on the first factor. Actually, this point can be misleading as will be explained later, but it appears intuitively unappealing to group uncorrelated variables together as measures of the same common factor. Thirdly, with so many variables having high loadings on a factor, interpretation is made difficult. The situation may be like trying to describe a man in terms of such disparate variables as weight, extraversion, intelligence, income, sense of humour, etc.; finding a generic term to cover all of these is virtually impossible. Nevertheless, in some cases the interpretation of a general factor may be feasible. Little more can be said about this first factor at present.

As far as the second factor is concerned, although all the variables again have significant loadings, some are positive and some are negative; this is generally known as a bipolar factor. Bipolar factors are often interpreted in terms of two quite different aspects at opposite poles of the underlying factor. In this case, for instance, if we examine the positive loadings we see that they all appear to be associated with 'length', which suggests that one pole could be referred to as body 'length'. On the other hand, the negative loadings all appear to be associated with the quality of body mass, which implies that the other pole might be referred to as body 'bulk'. It may be thought confusing that two such apparently different attributes emerge as extremes of the same factor and why the factor analysis hasn't produced two distinct factors, one 'body length' and another 'body bulk'. The explanation is that these two attributes are not in fact independent but may be thought of as two extreme body types, i.e. the poles being a very lean type and a very thick-set type. Burt and Banks concluded in their original work that this result supported the idea that there were two types of body build. This interpretation of the second factor following the factor analysis would appear to substantiate the casual observations made in section 2.1, where it was noted that variables 1, 2 and 3 appeared to form one group and 6, 7, 8 and 9 another.

The bipolar nature of the loadings on the second factor (or subsequent ones) is a consequence of the centroid method being used here, and the bipolarity may disappear after rotation; hence we should not lean too heavily on this interpretation that there are two distinct body types. Since standing height appears at one extreme of this second factor as an indicator of body length, whilst abdomen girth appears at the other extreme as an indicator of body length, and since it appears that this factor reveals that there is a tendency for people to fall into one of two groups – the lean or the thick-set type – a number of points need to be made. One is, if thick-set people tend to have short body length, why doesn't standing height correlate negatively with abdomen girth? The answer to this is that the correlation between two variables depends upon the influence of other determinants (or factors) which they measure in common, i.e. factors F_1 and F_3 as well as F_2. This is explained fully below.

The correlations between variables and factors may be held to 'explain' or 'account for' the correlations between variables, in the sense that a functional relation exists between them. This is:

$$r_{ij} = \sum_k F_{ik} F_{jk}, \tag{2.1}$$

where r_{ij} is the correlation between variables i and j, and F_{ik}, F_{jk} represent the correlations between i and the kth factor and j and the kth factor respectively. Thus for variable 1 (standing height) and 6 (abdomen girth) the equation becomes:

$$r_{16} = \sum_{k=1}^{3} F_{1k} F_{6k} = F_{11} F_{61} + F_{12} F_{62} + F_{13} F_{63}$$

$$= (0.76 \times 0.63) + (0.50 \times -0.49) + (0.12 \times -0.29)$$

$$= 0.48 - 0.25 - 0.03 = 0.20.$$

From the second term in the above expression it can be seen that the bipolarity of F_2 does indeed tend to produce a negative correlation between variables 1 and 6 (i.e. $F_{12} F_{62}$), but the fact that both of them correlate positively (and highly) on F_1 hides this and the overall effect is to produce a positive correlation between variables 1 and 6. Thus, the fact that there is a possible distinction between body types is hidden in the original correlations and only emerges after the analysis of these into factors. The original variables are *multifactorial*, i.e. impure measures of several unobserved underlying variables. A second aim of factor analysis may therefore be said to aid theory construction where this is hindered by the existence of only unsatisfactory (i.e. impure) measures which may be relevant to such a theory.

The actual value of r_{16} can be seen to be 0.17 from Table 2.1 and not 0.20 as calculated using equation (2.1). The difference is not marked, however, and the three factors can be said to satisfactorily reproduce the original correlation.

The inclusion of further factors would reduce the difference between actual correlations and calculated values, but since this would result in a loss of parsimony, we are prepared to tolerate a certain degree of discrepancy.

When we discussed the first or general factor, the second objection we raised was that the factor was made up of some variables (e.g. 4 and 8) which appear to have nothing in common. This appears to defy commonsense until we recognise the multifactorial nature of each variable. Applying equation (2.1) to these variables we obtain

$$r_{48} = (0.54 \times 0.56) + (0.48 \times -0.43) + (-0.43 \times 0.13) = 0.04.$$

Hence it is entirely possible for variables 4 and 8 to be measures of a general factor whilst at the same time being virtually uncorrelated.

The third factor has only four significant loadings, two positive and two negative. Following the above reasoning, we may interpret one pole as any measurement above the waist and the other measurement below. This suggests a difference between trunk length and leg length, i.e. that there is a slight tendency for people to have either long dimensions above the waist or long dimensions below.

It should be reiterated that the interpretations above must be regarded as very tentative – they represent only one way of looking at the data. This will emerge more clearly after rotation has been discussed.

In devising names for a factor, we have made use of variables which have relatively high loadings on the factor. This is a dangerous procedure, particularly when factor loadings are low. A factor loading is a correlation between a variable and a factor, and even one as high as 0.5 means that only 25 per cent (F_{ik}^2) of the variance of one is explicable from the other. The other 75 per cent of the variance of a variable may be determined by the very thing that causes us to give that variable its name, which we may then erroneously use to interpret the factor. It is advisable in any factor analytic study to be conservative about interpretation and to make it part of a programme of study in which a gradual picture is built up of the meaning of the factors. Many variables which the investigator believes to be related to particular factors should be included in the analysis to ensure over-determination of the factor. There is no hard and fast rule about this, but a rule of thumb is to include five times as many variables as expected factors.

Having briefly examined the results of the factor analysis, we will now proceed with a detailed explanation of the centroid method of factor analysis of the problem discussed above.

2.3 The centroid method: extraction of the first factor

We start with the correlation matrix of Table 2.1, reproduced in Table 2.3.

Notice that the top right-hand part of the matrix has been completed and that the values on the principal diagonal have been filled in – these values are known as 'communalities' and are normally designated 'h^2'. The total variance associated with each variable may be thought of as having two components: that variance which it shares with other variables (known as the 'common variance'), and the variance specific to itself (known as the 'unique variance'), although the latter

TABLE 2.3 *Correlation matrix for the nine body dimensions: extraction of the first factor*

Variable	1	2	3	4	5	6	7	8	9
1	*0.81*	0.81	0.59	0.58	0.53	0.17	0.33	0.22	0.40
2	0.81	*0.81*	0.67	0.29	0.25	0.13	0.39	0.29	0.41
3	0.59	0.67	*0.67*	0.19	0.16	0.08	0.28	0.21	0.29
4	0.58	0.29	0.19	*0.67*	0.67	0.23	0.17	0.08	0.28
5	0.53	0.25	0.16	0.67	*0.67*	0.29	0.22	0.12	0.33
6	0.17	0.13	0.08	0.23	0.29	*0.77*	0.70	0.52	0.77
7	0.33	0.39	0.28	0.17	0.22	0.70	*0.83*	0.59	0.83
8	0.22	0.29	0.21	0.08	0.12	0.52	0.59	*0.62*	0.62
9	0.40	0.41	0.29	0.28	0.33	0.77	0.83	0.62	*0.83*

also contains an amount of error variance. The fact that such common variance exists means that those variables which share common variance in part measure the same things, i.e. common factors. The extent to which a variable possesses unique variance is an indication that this variable measures something which none of the other variables in the set measure.

It was pointed out in Chapter 1 that r_{ij}^2 is a measure of the shared variance or common variance between variables i and j; it is the extent to which the variance of variable i is explicable from or accounted for variable j. Ignoring error of measurement for the moment, r_{ii} should be unity since a variance would be expected to correlate perfectly with itself. If we apply equation (2.1) to r_{ii}, this will in general be less than one since we are taking account only of factor loadings, i.e. correlations between variable i and factors which are common to variable i and other variables in the set under consideration. Other factors, unique to variable i, are not taken into consideration. Thus the entries in the leading diagonal are not referred to as r_{ii} (or r_i^2) but as h_i^2. This is the extent to which the variance of variable i is accounted for by common variance; $(1 - h_i^2)$ is the remainder, i.e. unique variance.

Thus, for three common factors, equation (2.1) gives us

$$h_i^2 = \sum_k F_{ik}^2 = F_{i1}^2 + F_{i2}^2 + F_{i3}^2, \tag{2.2}$$

e.g.

$$h_i^2 = 0.76^2 + 0.50^2 + 0.12^2 = 0.84.$$

Communality is the measure of common variance. Before carrying out a factor analysis it is necessary with some methods (including the centroid method), to fill in the principal diagonal of the correlation matrix with suitable values for the communalities. As we have seen from equation (2.2), it is easy to calculate such values once the factor loadings are known. But these are not known prior to carrying out the factor analysis. How then can we estimate values for the communalities at the outset?

Several methods have been proposed and used but none is markedly superior to another. Thurstone [1947] has suggested a simple method which amounts to choosing as the communality value the highest correlation which the variable in question has with any one of the other variables in the set; this method has been employed in the following example. The highest correlation which variable 1 enjoys with the other variables is 0.81 with variable 2, thus 0.81 is chosen as the common variance or communality value for variable 1 and is entered in the first position in the principal diagonal. In the same way, 0.81 is chosen as the second communality, 0.67 as the third and so on.

There are more complex methods of estimating the communalities, one of which will be illustrated in the next chapter. Further details concerning this problem can be found in Harman [1967]. After determining the communalities, we can proceed to the extraction of the first factor.

Let r_{ij} refer to the elements of the correlation matrix (i refers to rows and j to columns). We first sum each column to obtain $\sum_i r_{ij}$, then add these sums to give sum of sums $\sum_{ij} r_{ij}$. This gives Table 2.4.

TABLE 2.4 *Column sums* $\sum_i r_{ij}$ and sums of sums $\sum_{ij} r_{ij}$

	Columns/variables									sum of sums
	1	2	3	4	5	6	7	8	9	
Column sums	4.44	4.05	3.14	3.16	3.24	3.66	4.34	3.27	4.76	34.06

For example:

$$\sum_i r_{i1} = 0.81 + 0.81 + \ldots + 0.40 = 4.44,$$

and sum of sums,

$$\sum_{ij} r_{ij} = 4.44 + 4.05 + \ldots 4.76 = 34.06.$$

The first factor loadings are then obtained from the formula:

$$F_{j1} = \sum_i r_{ij} / \sqrt{\sum_{ij} r_{ij}}, \tag{2.3}$$

where F_{j1} is the loading of the jth variable on the first factor. F_{11}, the factor loading of the first variable, is equal to, for example:

$$4.44/\sqrt{34.06} = 0.76.$$

The factor loadings on the first factor are shown below in Table 2.5.

TABLE 2.5 *First factor loadings*

	Column/variables								
	1	2	3	4	5	6	7	8	9
F_{j1}	0.76	0.69	0.54	0.54	0.56	0.63	0.74	0.56	0.82

2.4 Expected values after computation of first factor

From equation (2.1) it follows that if the original correlations are explicable in terms of only one factor (the first), then:

$$r_{ij} = F_{i1} F_{j1} \tag{2.4}$$

approximately (allowing for error), since factor loadings on second and subsequent factors will be zero. In order to test whether this is the case we compute the matrix we would expect to obtain if only one factor is present; this is referred to as the expected values matrix, ${}_1E_{ij}$ (Table 2.6). The elements of this first expected values matrix ${}_1e_{ij}$ are computed from:

$${}_1e_{ij} = F_{i1} F_{j1} \tag{2.5}$$

TABLE 2.6 *First expected values* ${}_1E_{ij}$

Variable	1	2	3	4	5	6	7	8	9
1	0.58								
2	0.52	0.48							
3	0.41	0.37	0.29						
4	0.41	0.37	0.29	0.29					
5	0.43	0.39	0.30	0.30	0.31				
6	0.48	0.43	0.34	0.34	0.35	0.40			
7	0.56	0.51	0.40	0.40	0.41	0.47	0.55		
8	0.43	0.39	0.30	0.30	0.31	0.35	0.41	0.31	
9	0.62	0.56	0.44	0.44	0.46	0.52	0.61	0.46	0.67

For example:

$${}_1e_{23} = F_{21} F_{31} = 0.69 \times 0.54 = 0.37,$$

and so on.

If only one factor is present then ${}_1e_{ij}$ should equal r_{ij}, and to test whether this is in fact the case, we subtract the matrix of expected values from the original correlation matrix, and the resulting matrix, referred to as the first residual matrix, ${}_1S_{ij}$, is given below in Table 2.7.

TABLE 2.7 First residual values ${}_1S_{ij}$

Variable	1	2	3	4	5	6	7	8	9
1	*0.23*	0.29	0.18	0.17	0.10	−0.31	−0.23	−0.21	−0.22
2	0.29	*0.33*	0.33	0.30	−0.08	−0.14	−0.30	−0.12	−0.10
3	0.18	0.30	*0.38*	−0.10	−0.14	−0.26	−0.12	−0.01	−0.15
4	0.17	−0.08	−0.10	*0.38*	0.37	−0.11	−0.23	−0.22	−0.16
5	0.10	0.14	−0.14	0.37	*0.36*	−0.06	−0.19	−0.19	−0.13
6	−0.31	−0.30	−0.26	−0.11	−0.06	*0.37*	0.23	0.17	0.25
7	−0.23	−0.12	−0.12	−0.23	−0.19	0.23	*0.28*	0.18	0.22
8	−0.21	−0.10	−0.09	−0.22	−0.19	0.17	0.18	*0.31*	0.16
9	−0.22	−0.15	−0.15	−0.16	−0.13	0.25	0.22	0.16	*0.16*

If ${}_1s_{ij}$ is an element in this first residual matrix, then:

$${}_1s_{ij} = r_{ij} - {}_1e_{ij},$$

for example:

$$\begin{aligned}{}_1s_{23} &= r_{23} - {}_1e_{23}\\ &= 0.67 - 0.37 = 0.30\end{aligned}$$

We have stated above that if only one factor is present then all the ${}_1s_{ij}$ values would be zero (allowing for error), which is clearly not the case and suggests the presence of at least one other factor.

We therefore proceed to test the hypothesis that no more than two factors are present, which we can do by examining the equality:

$${}_1s_{ij} = F_{i2}F_{j2},$$

following on from equation (2.1).

If the second factor accounts for all of the residual element, then we can conclude that only two factors are present.

Before proceeding with this discussion a brief digression is in order. The square of a factor loading is the amount of variance of a variable attributable to or explicable by that factor. Thus the sum of squared first factor loadings is the amount of the total variance attributable to that factor. The higher the factor loadings, therefore, the greater the amount of variance explained.

The aim of factor analysis is ideally to explain as much of the original common variance as possible with each successive factor, i.e. to account for as much variance as possible with the fewest number of factors. The most efficient solution to this problem is achieved using the method of principal factor analysis, which we will discuss in Chapter 4. The centroid method is a simplification and an approximation of the principal factor procedure. The aim is thus to maximise the sum of each column in the correlation matrix, since this will increase the size of factor loadings (from equation (2.3)), by reducing the number of negative correlations to a minimum. This is achieved using a process called 'reflection', which amounts to reversing the scales of certain variables such that the sign of the correlations between those variables and other non-reflected variables changes, although the magnitude of the correlation remains unaffected.

The difficulty lies in an efficient choice of variables to be reflected so as to minimise the number of negative correlations in Table 2.7 and hence maximise the column sums. The experienced researcher will learn which are the best variables to reflect so as to achieve a quick solution. However, we present here a systematic procedure for efficient reflection which overcomes any difficulties of expertise.

2.5 Determination of reflected variables

Referring to Table 2.7 of first residual values, the method is as follows. Ignore entries in the principal diagonal, which are unaffected by reflection since they are correlations of a variable with itself and therefore necessarily positive. Next, let A_j represent the sum of positive entries and B_j the sum of negative entries in column j. It can be shown that reflecting that variable whose column has the largest value of $(B_j - A_j)$ maximises the increase in aggregate column sums. For example, the positive sums A_j and negative sums B_j, together with the $B_j - A_j$ values for Table 2.7, are shown below in Table 2.8.

TABLE 2.8 *Values of A_j and B_j*

	Variable								
	1	2	3	4	5	6	7	8	9
A_j	0.74	0.59	0.51	0.84	0.47	0.65	0.63	0.51	0.63
B_j	0.97	0.79	0.86	0.82	0.79	0.88	1.07	0.83	0.76
$B_j - A_j$	0.23	0.20	0.35	−0.02	0.32	0.23	0.44	0.32	0.13

The largest $(B_j - A_j)$ value is 0.44 in column 7. We now change the signs of all the correlations in Table 2.7, except principal diagonal values, involving variable 7. This produces the following changes in Table 2.7 shown in Table 2.9 opposite.

TABLE 2.9 *First residual matrix after reflecting variable 7*

Variable	1	2	3	4	5	6	7	8	9
1	*0.23*	0.29	0.18	0.17	0.10	−0.31	0.23	−0.21	−0.22
2	0.29	*0.33*	0.33	0.30	−0.08	−0.14	0.30	−0.12	−0.10
3	0.18	0.30	*0.38*	−0.10	−0.14	−0.26	0.12	−0.09	−0.15
4	0.17	−0.08	−0.10	*0.38*	0.37	−0.11	0.23	−0.22	−0.16
5	0.10	−0.14	−0.14	0.37	*0.36*	−0.06	0.19	−0.19	−0.13
6	−0.31	−0.30	−0.26	−0.11	−0.06	*0.37*	−0.23	0.17	0.25
7	0.23	0.12	−0.12	−0.23	−0.19	0.23	*0.28*	0.18	0.22
8	−0.21	−0.10	−0.09	−0.22	−0.19	0.17	−0.18	*0.31*	0.16
9	−0.22	−0.15	−0.15	−0.16	−0.13	0.25	−0.22	0.16	*0.16*

The $(B_j - A_j)$ values are computed again, and the procedure is successively repeated, variables being reflected until all the $(B_j - A_j)$ values are negative or until $\Sigma(B_j - A_j)$ can be reduced no further. In this example, variables 6, 9 and 8 must be reflected and the complete process is summarised below in Table 2.10; notice that all the $(B_j - A_j)$ values on the bottom row of this table are negative.

TABLE 2.10 *Summary of variable reflection procedure*

	Variable								
	1	2	3	4	5	6	7	8	9
A_j	0.74	0.59	0.51	0.84	0.47	0.65	0.63	0.51	0.63
B_j	0.97	0.79	0.86	0.82	0.79	0.88	1.07	0.83	0.76
$B_j - A_j$	0.23	0.20	0.35	−0.02	0.32	0.23	0.44	0.32	0.13
	Reflect 7								
$B_j - A_j$	−0.23	−0.04	0.11	−0.48	−0.06	0.69	−0.44	0.68	0.57
	Reflect 6								
$B_j - A_j$	−0.85	−0.64	−0.41	−0.70	−0.18	−0.69	−0.90	1.02	1.07
	Reflect 9								
$B_j - A_j$	−1.29	−0.94	−0.71	−1.02	−0.44	−1.19	−1.34	1.34	−1.07
	Reflect 8								
$B_j - A_j$	−1.71	−1.24	−0.89	−1.46	−0.82	−1.53	−1.70	−1.34	−1.39

2.6 Extraction of the second factor

After the reflection of variables 7, 6, 9 and 8, and the consequent sign changes, the first residual matrix appears as shown below in Table 2.11.

TABLE 2.11 *First residual matrix after complete reflections*

Variable	1	2	3	4	5	6	7	8	9
1	*0.23*	0.29	0.18	0.17	0.10	0.31	0.23	0.21	0.22
2	0.29	*0.33*	0.33	0.30	−0.08	0.14	0.30	0.12	0.10
3	0.18	0.30	*0.38*	−0.10	−0.14	0.26	0.12	0.09	0.15
4	0.17	−0.08	−0.10	*0.38*	0.37	0.11	0.23	0.22	0.16
5	0.10	−0.14	−0.14	0.37	*0.36*	0.06	0.19	0.19	0.13
6	0.31	0.30	0.26	0.11	0.06	*0.37*	0.23	0.17	0.25
7	0.23	0.12	−0.12	−0.23	−0.19	−0.23	*0.28*	0.18	0.22
8	0.21	0.10	0.09	0.22	0.19	0.17	0.18	0.31	0.16
9	0.22	0.15	0.15	0.16	0.13	0.25	0.22	0.16	0.16

The procedure for extracting the second factor is as before. We first sum the columns to give

$$\sum_{i} {}_1s'_{ij} \text{ and the } \sum_{ij} {}_1s'_{ij}$$

(where the dash indicates that the reflection process has been applied) which are given in Table 2.12.

TABLE 2.12 *Column sums $\Sigma_j s'_{ij}$*

	Variable									
	1	2	3	4	5	6	7	8	9	sum of sums
$\Sigma_1 s'_{ij}$ Column sums	1.94	1.57	1.27	1.84	1.18	1.90	1.98	1.65	1.55	14.88

The second factor loadings are then determined using the following formula (analogous to equation (2.3)):

$$F_{j2} = \sum_{i} {}_1s'_{ij} / \sqrt{\sum_{ij} {}_1s'_{ij}},$$

where F_{j2} is the loading on the jth variable of the second factor.

For example, F_{12}, the second factor loading on the first variable, is given by

$$F_{12} = 1.94/\sqrt{14.88} = 0.50.$$

The second factor loadings of the nine variables are shown below in Table 2.13.

TABLE 2.13 *Second factor loadings*

	Variable								
	1	2	3	4	5	6	7	8	9
F_{j2}	0.50	0.41	0.33	0.48	0.31	− 0.49	− 0.51	− 0.43	− 0.40

Note that the factor loadings corresponding to those variables which have undergone reflection have been assigned a negative value; this is done to reverse the effect of reflection. If the scale of these reflected variables had not been reversed, then any variable (including a factor) would have correlated negatively with them.

2.7 Expected values after extraction of second factor

If our hypothesis that there were only two factors was tenable, then we would expect that ${}_1s_{ij}$ before reflection would be approximately equal to $F_{12} F_{j2}$. In order to test this hypothesis, a table of expected values (Table 2.14) was constructed in a similar way to that which followed extraction of the first factor. If the elements of this matrix are designated ${}_2e_{ij}$, then they are computed from:

$$ {}_2e_{ij} = F_{i2} F_{j2}. $$

TABLE 2.14 *Second expected values* ${}_2e_{ij}$

Variable	1	2	3	4	5	6	7	8	9
1	0.25								
2	0.20	0.17							
3	0.17	0.13	0.11						
4	0.24	0.20	0.16	0.23					
5	0.15	0.13	0.10	0.15	0.10				
6	− 0.25	− 0.20	− 0.16	− 0.24	− 0.15	0.24			
7	− 0.26	− 0.21	− 0.17	− 0.24	− 0.16	0.25	0.26		
8	− 0.21	− 0.18	− 0.14	− 0.21	− 0.13	0.21	0.22	0.18	
9	− 0.20	− 0.16	− 0.13	− 0.19	− 0.12	0.20	0.20	0.17	0.16

In a similar way to before, the hypothesis that ${}_1s_{ij} = {}_2e_{ij}$ is tested by subtracting the elements of Table 2.14 from those of the first residual matrix, Table 2.7, to produce a table of second residuals, Table 2.15.

It is clear from inspection of Table 2.15 that the second hypothesis is not supported and that there is at least a third factor involved. If the elements of this

TABLE 2.15 *Second residuals*

Variable	1	2	3	4	5	6	7	8	9
1	*– 0.02*	0.09	0.01	– 0.07	– 0.05	– 0.06	0.03	0.0	0.02
2	0.09	*0.16*	0.17	– 0.28	– 0.27	– 0.10	0.09	0.08	0.01
3	0.01	0.17	*0.27*	– 0.26	– 0.24	– 0.10	0.05	0.05	– 0.02
4	– 0.07	– 0.28	– 0.26	*0.15*	0.22	0.15	0.01	– 0.01	0.03
5	0.05	– 0.27	– 0.24	0.22	*0.26*	0.09	– 0.03	– 0.06	– 0.01
6	– 0.06	– 0.10	– 0.10	0.15	0.09	*0.13*	– 0.02	– 0.04	0.05
7	0.03	0.09	0.05	0.01	– 0.03	– 0.02	*0.02*	– 0.04	0.02
8	0.00	0.08	0.05	– 0.01	– 0.06	– 0.04	– 0.04	*0.13*	– 0.01
9	– 0.02	0.01	– 0.02	0.03	– 0.01	0.05	0.02	– 0.01	*0.00*

second residual matrix are designated ${}_2s_{ij}$, then we proceed to test the hypothesis that these residual correlations can be accounted for by a third factor, i.e. that

$$ {}_2s_{ij} = F_{i3}F_{j3}. $$

One need not be alarmed to find a negative value for one of the communalities (${}_2s_{11} = -0.02$); this merely means that the original estimate was too low. Since it is a small value, it is not necessary to begin again with a new estimate, but it can be regarded as zero or the sign changed. Another procedure which is sometimes adopted is to estimate the communalities afresh, after the extraction of each factor, by making each equal to the largest correlation in its column.

Variables 5, 4, 6 and 9 were reflected using the procedure described above. The relevant calculations are set out in Table 2.16.

TABLE 2.16 *Reflections of variables, 5, 4, 6 and 9*

	1	2	3	4	5	6	7	8	9
A_j	0.13	0.44	0.28	0.41	0.31	0.29	0.20	0.13	0.13
B_j	0.20	0.65	0.62	0.55	0.66	0.32	0.09	0.16	0.04
$B_j - A_j$	0.07	0.21	0.34	0.14	0.35	0.03	– 0.11	0.03	– 0.09
Reflect 5									
$B_j - A_j$	– 0.03	– 0.33	– 0.14	0.58	– 0.35	0.21	– 0.17	– 0.09	– 0.11
Reflect 4									
$B_j - A_j$	– 0.17	– 0.89	– 0.66	– 0.58	– 0.79	0.51	– 0.15	– 0.11	– 0.05
Reflect 6									
$B_j - A_j$	– 0.29	– 1.09	– 0.86	– 0.88	– 0.97	– 0.51	– 0.19	– 0.19	0.05
Reflect 9									
$B_j - A_j$	– 0.33	1.07	– 0.94	– 0.94	– 0.95	– 0.61	– 0.15	– 0.21	– 0.05

2.8 Extraction of the third factor

Extraction of the third factor proceeds in exactly the same manner as for the first and second factors. The third factor loadings are shown in Table 2.17.

TABLE 2.17 *Third factor loadings*

	Variable								
	1	2	3	4	5	6	7	8	9
F_{j3}	0.12	0.49	0.48	− 0.43	− 0.48	− 0.29	0.07	0.13	− 0.02

Expected values were calculated as before and a table of third residuals produced, shown below in Table 2.18.

TABLE 2.18 *Third residuals*

	Variables								
	1	2	3	4	5	6	7	8	9
1	− 0.03								
2	0.03	− 0.08							
3	− 0.05	− 0.07	0.04						
4	− 0.02	− 0.07	− 0.05	− 0.03					
5	0.01	− 0.03	− 0.01	0.01	0.03				
6	− 0.03	0.04	0.04	0.03	− 0.05	0.05			
7	0.02	0.06	0.02	0.04	0	0	0.01		
8	− 0.02	0.02	− 0.01	0.04	0	0	− 0.05	0.09	
9	− 0.02	0.01	− 0.02	0.03	− 0.01	0.05	0.02	− 0.01	0

These do not appear to be markedly different from zero and thus the third hypothesis is justified, i.e. three factors are sufficient to explain the original correlations between the set of variables. This rule of thumb as to the adequacy of the number of factors to be extracted is no more than that. Some maximum value which the elements in the final residual should not exceed is determined before the analysis commences, and when the residual elements are smaller than this value, the analysis can stop. Whilst this criterion for stopping the analysis may be perfectly satisfactory for the more experienced worker, it may be of some interest to present a more rigorous rule. This is due originally to Guttman [1954] and developed by Kaiser, and the subject is dealt with in some more detail in Harman [1967]. Kaiser's criterion is that the variance accounted for by each factor, i.e. the sum of the squared loadings on each factor, should exceed 1.0.

If the scores on each variable are standardised, then each variable has unit variance. Kaiser's criterion ensures that a factor accounts for at least as much variance as a single variable. Another criterion, one which we will use here, is to consider only those factors which account for more than 10 per cent of the total variance. Since we are considering standardised scores on each variable, so that each variable has unit variance, the total variance of the set is equal to the number of variables. There is no need to actually go to the trouble of standardising each variable before conducting a factor analysis when product-moment correlations are used as a starting-point since the value of a correlation coefficient is unaffected by linear transformations of scale. Since the total amount of variance explained or 'extracted' by a factor is the sum of its squared loadings (also known as the 'latent root'), the proportion of variance attributed to the factor is simply,

$$\% \text{ variance extracted} = \frac{\text{latent root}}{\text{no. of variables}} \times 100.$$

Note that when the number of variables is small (less than 10), Kaiser's criterion is more demanding than the 10 per cent one, i.e. it is more likely to reject a factor as non-significant. When the number of variables is about 10, there is no difference between the two criteria, whilst as the number grows larger, Kaiser's criterion becomes less demanding, i.e. it would accept a factor as significant even though the proportion of total variance it explains falls below 10 per cent. It is probably better to extract too many factors rather than too few for reasons to be explained in the next chapter, and from the foregoing discussion it can be seen that in choosing the least demanding criterion of these two (there are a number of others which will not be discussed here), the number of variables must be taken into account.

The sum of squared loadings (latent root) on the first factor in our example is given by:

$$0.76^2 + 0.69^2 + \ldots + 0.82^2 = 3.87.$$

Hence the percentage variance extracted by the first factor is

$$\frac{3.87}{9} \times 100 = 43\%.$$

Similarly, the latent roots for F_2 and F_3 are 1.71 and 0.99 respectively; the percentage variances extracted are 19 and 11 per cent. Note that, strictly speaking, the third factor fails Kaiser's criterion, i.e. its latent root is 0.99, but it is sufficiently close to 1, especially in view of the small number of variables, to justify inclusion.

In general, the percentage variance extracted by successive factors drops, and whilst this is invariably the case with principal component analysis which is discussed in Chapter 4, it will not always be so with a centroid analysis, which can be thought of as an approximate method.

This concludes the straightforward factor analysis of this problem. The set of factor loadings on each variable were shown in Table 2.2, which also shows the communalities calculated from the sums of squared factor loadings for each variable. In our case, they are not much different from those estimated at the start of the analysis but, if desired, they could be used as revised estimates of the communalities and the entire factor analysis repeated. This can be done as often as desired. An iterative procedure for estimating communalities in this way will be employed in Chapter 4.

What has been achieved so far is usually considered to be only the first step in the analysis of relationships between the original set of variables. In the next chapter we will look at the problem of rotation, an important step in the process of interpreting the factors which have been extracted.

3 Rotation of Factors

3.1 Introduction: orthogonal rotation

The results of a factor analysis may be represented graphically as points in n-dimensional space where the factors represent the axes and the factor loadings the co-ordinates of each point. This is only conveniently realisable graphically when an analysis yields a two-factor solution. To illustrate this idea let us consider an analysis having five variables and giving the two-factor solution shown in Table 3.1

TABLE 3.1 *Factor matrix for five variables*

		Factor	
		F_1	F_2
Variables	1	0.6	0.3
	2	0.7	0.5
	3	0.4	0.4
	4	0.5	− 0.8
	5	0.8	− 0.4

These results can be represented graphically using the factor loadings as co-ordinates as in Figure 3.1.

The centroid method of analysis ensures that the first factor axis passes through the centroid (or average of the co-ordinates) of the points representing the variables, and that the second factor will be at right angles, i.e. orthogonal, to it. It is because of this that the first factor always appears a general one and the second factor always appears bipolar. Each additional factor will also be orthogonal to preceding ones and hence will be bipolar. The aim of the centroid method may thus be said to be to explain the observed correlations in terms of a set of axes, each of which is independent of the others. If the reader has any difficulty relating the geometric representation to the algebraic one, he should recall that the co-ordinates of a point in Fig. 3.1 are given by the factor loadings or correlations between variables and factors. A perfect measure of F_1 would have

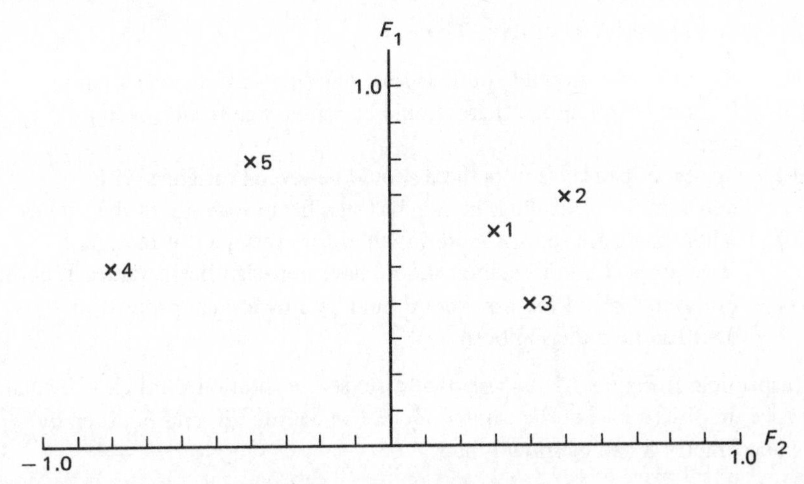

Fig. 3.1 *Graphical representation of the loadings on two factors of five variables*

co-ordinates (0, 1) and a perfect measure of F_2 would have co-ordinates (1, 0). These two measures would not correlate at all (cf. equation (2.1)), and hence F_1 and F_2 are independent. Points in the diagram which lie near to each other in space will correlate highly and positively whilst those at a distance from each other will not, or will correlate negatively.

The position of the axes is determined by the particular method of factor analysis employed and the imposition of this geometric structure may not be the most immediately amenable to interpretation of the factors.

For example, consider the fact that all the variables tend to correlate highly with the first factor. One might conclude from this that F_1 is a general attribute which all the variables tend to measure. But this may not be reasonable, as was described in the previous chapter, since the situation may be like trying to describe a person in terms of one attribute which is a composite of many heterogeneous variables which may have little in common with one another (e.g. age, sex, weight, height, intelligence etc.). The variables involved may in fact be *totally* uncorrelated with one another.

Since the general purpose of factor analysis is to discover underlying relationships between clusters of variables, i.e. variables sharing similar characteristics, this process of identification may be improved if the axes are re-drawn in such a way as to more easily distinguish between groups or clusters of variables which have something in common and which may then be seen to reveal common characteristics not previously perceived with the axes in the original constrained position. One solution to this problem of axis re-orientation is due to Thurstone [1947], who proposed 'rotation to simple structure', the effect of which is to maximise the number of loadings with negligible, i.e. non-significant, values, whilst preserving a few values with high loadings. This may be effected most simply by a rotation 'by eye', followed by a check of the revised factor matrix, after rotation, against five criteria. These are:

(i) Each variable should exhibit at least one (non-significant) loading.
(ii) If there are n factors, there should be n non-significant loadings in each factor.
(iii) For every pair of factors there should be several variables with non-significant loadings in one, but significant loadings in the other.
(iv) Where there are four or more variables for every pair of factors, a large proportion of loadings should have non-significant values in both.
(v) For every pair of factors there should be only a small proportion of significant loadings in both.

Inspection of Figure 3.1 suggests that a clockwise rotation through 45° might improve the situation. Let the rotated factors be denoted F'_1 and F'_2, then by simple geometry it can be shown that:

$$F'_1 = F_1 \cos\theta + F_2 \sin\theta$$

$$F'_2 = F_2 \cos\theta - F_1 \sin\theta,$$

where θ is the angle of rotation, in this case 45°. This produces the transformed factor matrix shown below in Table 3.2.

TABLE 3.2 *Revised factor matrix*

		Variable				
		1	2	3	4	5
Factor	F'_1	*0.64*	*0.85*	*0.57*	− 0.21	0.28
	F'_2	− 0.21	− 0.14	0.0	− *0.92*	− *0.85*

This rotation appears to satisfy the five criteria of Thurstone and thus appears to have been successful. The question of interpretation now presents itself. It seems that we have identified two distinct clusters of variables: one, two and three having high loadings on F'_1, and four and five having high loadings on F'_2. This distinction was not so obvious in the unrotated factor matrix and an inspection of the variables in each cluster for what they appear to have in common would thus enable us to more easily interpret the two factors.

With only two factors, the problem of rotation is straightforward since geometric display of the situation is simple. With three or more factors, however, the problem becomes rapidly more difficult; in this case one solution is to take the factors two at a time.

Rotation of the three-factor solution described in Chapter 2 yields the following revised factor matrix (Table 3.3). This follows two rotations, F_1 and F_2 through 45° and F_2 and F_3 through 45° (note that F''_2 indicates that this factor has been rotated twice).

TABLE 3.3 *Rotation of three-factor problem*

	Variable									% variance extracted
	1	2	3	4	5	6	7	8	9	
F_1'	– 0.18	– 0.20	– 0.15	– 0.04	– 0.18	– *0.79*	– *0.88*	– *0.70*	– *0.87*	32
F_2''	*0.54*	0.21	0.10	*0.81*	*0.78*	0.28	0.06	– 0.03	0.22	19
F_3	*0.71*	*0.90*	*0.78*	0.21	0.10	– 0.13	0.16	0.16	0.19	23
h^2	0.83	0.89	0.64	0.70	0.65	0.72	0.80	0.52	0.84	(73)

Reference to Table 3.3, in which the significant loadings are shown in italic, shows how rotation of the factors has caused the three separate variable clusters to emerge quite sharply and agree with our original guesses. If we compare these results with the unrotated factor matrix, Table 2.2, it can be seen that the separate clusters are much more easily defined.

It is now possible to re-interpret these factors following rotation. Factor one, corresponding to variables 6, 7, 8 and 9, might be referred to as 'body bulk'; factor two (ignoring variable 1 for the moment) has high loadings for variables 4 and 5 and may be referred to as 'leg length'; and factor three (corresponding to variables 1, 2 and 3) perhaps as 'trunk length'. Notice that although clusters have emerged rather clearly following rotation, variable 1 still has significant loadings on two factors, but this is perhaps not surprising since this variable, 'standing height', would be expected to relate to both leg and trunk length or, put another way, both these factors would be expected to contribute to standing height, which may be referred to as a hybrid variable.

It was pointed out in the previous chapter that, in the Burt and Banks study, one variable (standing height) was a linear combination of two of the others, and that this was a situation to be avoided in general. If A is a linear combination of B and C, and the latter two variables tend to load highly on different factors, A is quite likely to be a hybrid variable, making interpretation difficult, and attempts to make it conform to the 'simple structure' solution can distort the interpretation. Moreover, since A will very probably correlate with B or C or both, its inclusion may lead to the extraction of spurious factors.

The purpose of orthogonal rotation is to locate clusters of related variables which are relatively independent of (or orthogonal to) other clusters; the factor 'body bulk' is independent of factors concerned with body length. Our interpretation of factors can therefore alter considerably following rotation, e.g. there now appears to be no factor common to all the variables and since 'body bulk' is independent of factors concerned with 'body length', we are not led to conclude that there are two distinct types of body-build, which was the conclusion suggested by Burt and Banks based on their unrotated data.

Notice that, following orthogonal rotation, communalities for each variable remain unchanged (except for rounding errors), and although the total variance extracted by the three factors is the same, the amount due to each has become more evenly distributed. In addition, the original correlations are just as reproducible from the rotated factor loadings as they were from the original ones. Nothing has therefore changed except the position of the axes used and the factor loadings (or correlations between variables and factors).

The fact that rotation produces a more even distribution of the variance accounted for by each factor poses something of a problem. Suppose we find that the three factors extracted in a particular analysis account for 60, 20 and 5 per cent of the variance respectively. Using the 10 per cent criterion, we may reject the third factor as being unimportant. However, if we kept it in and then rotated the factors we may find that it has had a considerable amount of variance distributed to it through the rotation procedure. On the whole, it is better to extract too many factors than too few, and we therefore recommend Kaiser's criterion rather than the 10 per cent one when there are more than 10 variables in a study, since it allows more factors to be extracted. We shall return to this question in section 3.4, where we discuss analytical methods of rotation.

We can see from the above discussion that rotation can markedly affect our interpretation of factors. Without rotation, the nature of the analysis ensures that the first factor, which accounts for more variance than subsequent ones, is a general one indicating that all, or most, of the variables measure something in common. This may or may not be valid theoretically. For example, a psychologist investigating mental abilities may wish to know whether or not there is a general ability which is manifested by each variable (or test) he employs in his study, or whether intellectual ability consists of several relatively independent specific abilities. However, one must beware of concluding that there is a general factor from the analysis alone, since rotation can alter the picture considerably. Moreover, one should beware of factors that emerge spuriously, since they can affect our interpretation of other factors even after rotation. For example, in an economics investigation where each variable is measured in monetary units varying over time, e.g. X_1 = average income for 1950, 1951, 1952, . . ., 1973, all the variables will correlate because of changes in monetary value over the time period, and this will be reflected in the emergence of a common factor. The economist will, of course, be aware of this and concern himself with the other factors which emerge as being of more interest. However, after rotation, the variance associated with the spurious, time-based common factor will be distributed to the other factors, making them appear perhaps more important than they really are. To guard against this, it is important to ensure that unwanted or spurious sources of common variance are not allowed to intrude into the analysis, artificially inflating the initial correlation between variables.

From the mathematical point of view (i.e. being able to explain or reproduce the observed correlations between variables), one position for the axes is as good as any other but, since this position affects our interpretation of the meaning of the factors we have extracted, it is important for the researcher to know where to position the rotated axes. Since they can be rotated to an infinite number of positions, many of which have radically different interpretations, the reader may be wondering whether almost any theoretical idea a researcher holds about the interrelationships among the variables he is studying can be upheld by suitable rotation of the axes. The answer is definitely 'no' because, although the position of the axes can change ad infinitum, the configuration of points in the space defined by those axes cannot. A cluster of related variables which are close together in n-dimensional space remain so regardless of where the axes are placed.

Rotation to approximate simple structure aids interpretation of factors and, if our primary interest is merely parsimony of description, this can be sufficient. However, the aim of rotating to simple structure is to make variables approximate to pure factor measures, i.e. measures of one factor only. Often this is unreasonable, since we know that many variables are composite; an example is 'standing height', which is a measure of body length both above and below the waist. But rotation to simple structure is no guarantee that the most useful positioning of the axes has been achieved. As was noted in Chapter 1, one of the difficulties encountered by behavioural scientists is that the variables which they can measure directly are often impure ones, and factor analysis can help with this difficulty by producing purer, though indirectly measurable, variables for use in research. The test of whether this has been achieved by factor analysis can only be gauged by using factors in research and seeing whether stable generalisations can be inferred. The true test of the optimum position for a rotated axis is whether or not it leads to factors that enable advances in subsequent research to be made, not necessarily fulfilling the criteria for simple structure. As more studies are done in a particular area, and knowledge is gained about the possible factorial composition of variables, so the investigator can gradually acquire the knowledge necessary to position axes in the most useful way. This can sometimes be done after a programme of study where some variables are used again and again in combination with new ones. As knowledge is gained about the factorial composition of the old variables, it can be used to position axes and hence to infer the factorial composition of new ones. Thus in a new study the researcher would include old variables, the factor composition of which is known. One should not be too confident about the factors arising from a single study in isolation, simple structure or no simple structure. Nevertheless, as a first step in a relatively unexplored field, simplification and ease of interpretation are important first considerations. But as a field becomes better understood, the major considerations should be the testing of hypotheses and the agreement or disagreement with data from previous studies.

3.2 Oblique rotation

In the body-size example, rotation to approximate simple structure still left variable 1 with significant loadings on two factors. Can we improve the approximation to simple structure at all? Often it is possible to do so by making use of oblique rotation.

In orthogonal rotation the axes remain at right angles to each other, which means that the factors, represented by the axes, are necessarily uncorrelated. Since one of the purposes of rotation is to improve the identification of clusters of similar variables, the use of orthogonal axes imposes an undesirable and unnecessary constraint and may frequently impose a structure in which some variables have significant loadings on more than one factor, as in the body-size example. This makes interpretation of the factors more difficult. Relaxation of the orthogonal constraint is possible and if the axes can be located freely at oblique, i.e. non-orthogonal, angles, the analysis may be improved. Isolation of separate variable clusters and the subsequent interpretation of the oblique factors may be thus made easier. The process of oblique rotation requires the use of a computer for all but the most trivial problems.

Oblique rotation by hand of more than two factors is rather complicated, so we have chosen a simple example, which will be analysed in more detail in the following chapter, to illustrate the points here. In this example, five variables are factor analysed and two factors emerge. The factor loadings are reproduced in Table 3.4.

TABLE 3.4 *Factor matrix of five variables*

		F_1	F_2	h^2
Variable	1	*0.67*	*0.53*	0.71
	2	*0.63*	0.17	0.43
	3	*0.76*	− 0.13	0.59
	4	*0.67*	− 0.21	0.50
	5	*0.72*	− 0.31	0.62
% variance extracted		48	9	(57)

In this example, orthogonal rotation does not lead to simple structure, whilst oblique rotation does. If F_1 is rotated through 58°, and F_2 through 40°, as in Figure 3.2, the resulting factor loadings appear as in Table 3.5. The relevant equations for deriving revised factor loadings are:

$$F_1' = F_1 \cos\theta + F_2 \sin\theta$$

$$F_2' = F_2 \cos\alpha - F_1 \sin\alpha,$$

where θ and α are the angles through which F_1 and F_2 are rotated, respectively.

Following this oblique rotation, the sums of squares of factor loadings for each variable no longer give us the communalities, and the sum of squares in each column can no longer be used to determine the amount of variance extracted or attributable to each factor. How this can be done after oblique rotation will be discussed in a later chapter. Moreover, the factor loadings cannot be combined in ways described previously to reproduce the original correlations. The purpose of

TABLE 3.5 *Rotated factor matrix (oblique)*

		F_1'	F_2'
Variable	1	*0.81*	– 0.06
	2	*0.48*	– 0.29
	3	0.29	– *0.60*
	4	0.18	– *0.60*
	5	0.12	– *0.71*

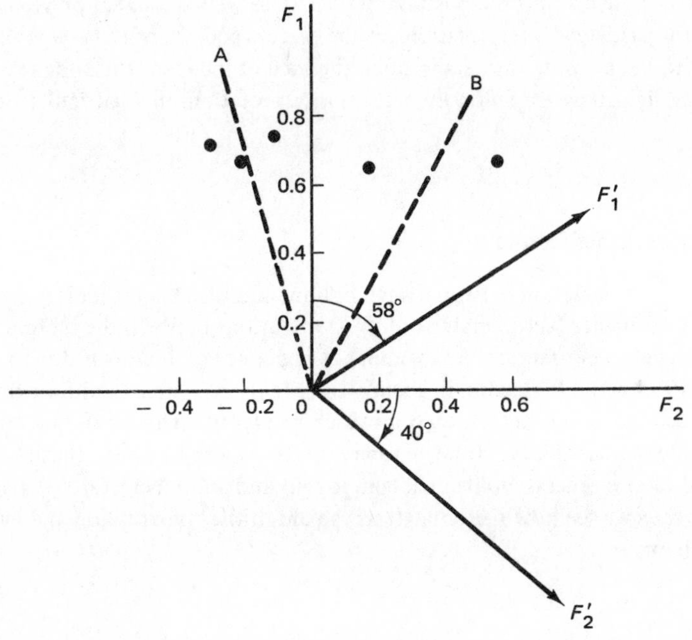

Fig. 3.2 *Oblique rotation of the two factors emerging from the factor analysis of five variables*

oblique rotation is simply to enable us to locate clusters of similar variables, and thus aid interpretation of factors.

The cosine of the angle between the axes representing factors is equal to the correlation between them; thus, for orthogonal axes, cos 90° = 0, indicating zero correlation between them. The smaller the angle between axes, the larger the cosine and, hence, the correlation. In the example we have given for oblique rotation, the correlation between F_1' and F_2' is given by cos 72° which is 0.31. How this affects the interpretation of the analysis had best be left until the next chapter, and all that need be said here is that oblique rotation has helped locate clusters of similar variables, and thus aided interpretation of the factors. We will also leave the oblique rotation of the three factors involved in our body-size example until Chapter 4.

One problem with oblique factors is that although they are useful for locating clusters of similar variables, they are themselves correlated, and if this correlation is substantial, then it may be necessary to introduce further underlying variables to explain the inter-correlations between factors, as described briefly in the next section; thus, if this is the case, the result is a loss of parsimony of description. It is better from this point of view if factors are not made too oblique, so that the correlations between them remain negligible. On the other hand, it may be that in a particular field of enquiry, such a loss of parsimony is necessary for building a satisfactory theory about the interrelationships between the original variables. The only final test about the position of axes is, as we have stated previously, how useful particular interpretations of the factors and relations between them turn out to be, i.e. how they stand up in the face of independent evidence and how much light they throw on theoretical interpretations in a particular field of enquiry.

3.3 Higher-order factors

The fact that correlation exists between oblique factors means that it is possible to carry out further factor analysis of the correlations between the factors and extract higher-order factors. An example of higher-order factors is due to Cattell [1963 and 1967] who, in a study on intelligence, extracted four first-order and two second-order factors, each of which he identified as a different type of general intellectual ability. Hitherto, one popular theory had been that ability consisted of *one* general intellectual component and a number of more specific abilities; thus we see how factor analysis can aid in the construction and evaluation of theory.

3.4 Analytical methods of rotation

As we mentioned in Chapter 1, the researcher would never normally carry out a factor analysis by hand but would leave the computation to a computer. It is also true that both orthogonal and oblique rotations can be left to a computer. There are a variety of analytic methods for which computer programs are available, some of which will be briefly described in this section.

One method, called Quartimax, attempts to achieve simple structure by making the rows of the factor matrix all 0's except one. This method does not distribute the variance away from the first factor sufficiently and therefore it tends to give a general factor which can be difficult to interpret.

A more widely-used method, known as Varimax, attempts to push up high loadings on a factor and reduce small loadings on the factor to zero, i.e. it attempts to maximise the variance of factor loadings for each factor. This method distributes the variance more equally than Quartimax, but excessively so, in that if unimportant factors are not excluded before the rotation procedure, they will become important by having some of the variance distributed to them.

Another method, proposed by Comrey [1973] and known as the Tandem method, also distributes the variance more evenly but not so much as Varimax, and therefore unimportant factors are not so prone to have spurious significance attached to them; in fact, this method allows us to reject unimportant factors during the rotation procedure. The Tandem method employs two criteria for rotation in succession. The first is that variables which correlate appear on the same factor. Factors which account for little variance after this criterion is applied are dropped from the analysis. The second criterion is that uncorrelated variables do not appear on the same factor. This method gives us an additional way of determining the number of common factors besides Kaiser's and the 10 per cent variance criteria described earlier.

The analytical methods described above are all concerned with orthogonal rotations. Similar methods, employing rather similar criteria, are available for achieving approximations to simple structure via oblique rotations; one such method, known as Promax, will be employed in the following chapter. Really satisfactory analytical oblique solutions have not yet been achieved because, although simple structure may be attained, there is still the problem of determining how oblique the factors should be allowed to go.

4 Principal Factor Analysis

4.1 Introduction

In Chapter 2 we examined the method of factor analysis using the centroid technique. It was stated then that the aim of the analysis was to explain the correlations between the original observed *variables* in terms of their correlations with a smaller set of *factors*. In this chapter we will examine two other methods of analysis, principal factor analysis and principal component analysis. The approach of the two methods is similar and their aim, to aid interpretation of the underlying structure of the interrelationships between variables, is the same. But there is in fact, as we shall see later, a fundamental difference between the two methods.

Principal factor analysis is similar in aim to the centroid method. In fact, the latter may be regarded as a rough approximation to the former; both seek to explain the inter-correlations between variables with as small a number of common factors as possible. That is, if the communality represents the common variance of each variable with all other variables (that due to common factors), then factor analysis seeks to analyse this common variance. The important difference between the centroid method of factor analysis and principal factor analysis is that the latter extracts the maximum amount of variance due to each factor at each stage, whereas the centroid method only achieves this approximately; in other words, it is less efficient.

Principal component analysis is similar to principal factor analysis in that the maximum amount of variance is extracted at each stage. The difference is that instead of estimating (by whatever method) the values of the communalities (the terms in the principal diagonal of the correlation matrix), the entries are taken as unity. These unit values represent the total variance of each variable and the object of principal component analysis is, therefore, to explain the total variance, i.e. common variance plus unique variance, due to each variable, with as small a number of components as possible (unlike factor analysis which, as we have seen, seeks to explain only the common variance). It is as though we were treating all the variance due to each variable as common variance.

As a consequence, some of the variance extracted at each stage is unique variance. This 'contamination' is not too serious from the factor analytic point of view in the first few factors extracted (the proportion of common variance extracted decreases with each factor), although the contribution of unique variance to later factors (and its extraction) becomes more serious. Whether or not we get different results from using unities or communality estimates in the principal diagonal depends on the amount of common variance that exists between the original variables, as we shall see later.

It may be of interest to compare the centroid solution and a principal component's solution to the body-size problem of the previous chapter.

Table 4.1 shows that the two solutions are essentially the same. Putting 1's in the principal diagonal instead of communality estimates has not greatly changed the solution obtained. It is of some interest that the principal components solution has extracted more of the variance at each stage.

In the previous chapter, the centroid factors were rotated orthogonally by hand until approximate simple structure was achieved. In the principal component analysis, rotation with the Varimax method was incorporated in the computer program.

TABLE 4.1 *Comparison of centroid and principal components solutions to the body-size problem (significant loadings in italic)*

Variable	Centroid method			Principal components method		
	F_1	F_2	F_3	C_1	C_2	C_3
1	*0.76*	*0.50*	0.12	*0.75*	*0.56*	0.07
2	*0.69*	*0.41*	*0.49*	*0.70*	*0.42*	*0.44*
3	*0.54*	*0.33*	*0.48*	*0.56*	*0.40*	*0.53*
4	*0.54*	*0.48*	*− 0.43*	*0.54*	*0.44*	*− 0.58*
5	*0.56*	*0.31*	*− 0.48*	*0.56*	*0.34*	*− 0.62*
6	*0.63*	*− 0.49*	− 0.29	*0.66*	*− 0.57*	− 0.23
7	*0.74*	*− 0.51*	0.07	*0.78*	*− 0.46*	0.09
8	*0.56*	*− 0.43*	0.13	*0.62*	*− 0.47*	0.17
9	*0.82*	*− 0.40*	− 0.02	*0.84*	*− 0.41*	− 0.02
% variance extracted	43	19	11	46	21	15
Total % variance extracted		73			82	

The two orthogonally rotated solutions are given in Table 4.2 for comparison.

The results are very similar, and the Varimax procedure still leaves variable 1 (standing height) with significant loadings on two factors, making interpretation difficult. The fact that the loadings on F_1' and C_1', although similar in magnitude,

TABLE 4.2 *Comparison of orthogonally rotated factors for the centroid and principal component solutions to the body-size problem*

Factor		Variables								
		1	2	3	4	5	6	7	8	9
Hand-rotated centroid factor (from Table 2.20)	F_1'	−0.18	−0.20	−0.15	−0.04	−0.18	*−0.79*	*−0.88*	*−0.70*	*−0.79*
	F_2'	*0.54*	0.21	0.10	*0.81*	*0.78*	0.28	0.06	−0.03	0.22
	F_3'	*0.71*	*0.90*	*0.78*	0.21	0.10	−0.13	0.16	0.16	0.19
Varimax rotation of principal component	C_1'	0.14	0.20	0.11	0.07	0.15	*0.87*	*0.87*	*0.77*	*0.89*
	C_2'	*−0.53*	−0.16	−0.01	*−0.89*	*−0.88*	−0.22	−0.07	0.06	−0.21
	C_3'	*0.77*	*0.90*	*0.87*	0.17	0.10	−0.10	0.23	0.18	0.22

are of different sign (and the same for F_2' and C_2'), does not affect the interpretation of the factors. It simply means that the polarity of the axes has been reversed. In the previous chapter it was mentioned that there is no need for the axes representing factors to be orthogonal, and it was shown in a simple example how oblique rotations can aid interpretation of factors by giving a better approximation to simple structure. The principal components in the present example were rotated

TABLE 4.3(*a*) *Oblique rotated Promax principal components (from Table 4.1)*

		Variable								
		1	2	3	4	5	6	7	8	9
Oblique factor	C_1'	−0.03	0.06	−0.01	−0.06	0.04	*0.91*	*0.88*	*0.79*	*0.88*
	C_2'	*−0.41*	0.02	0.16	*−0.92*	*−0.92*	−0.16	0.06	0.19	−0.08
	C_3'	*0.72*	*0.92*	*0.92*	0.02	−0.07	−0.26	−0.12	0.11	0.08

TABLE 4.3(*b*) *Correlations between the oblique rotated factors*

	C_1'	C_2'	C_3'
C_1'	1.00	0.32	−0.29
C_2'		1.00	−0.37
C_3'			1.00

using the Promax procedure (which was carried out by computer), and the results are given in Table 4.3(*a*) along with the inter-correlations between components (factors), Table 4.3(*b*).

Table 4.3(*a*) shows that although the loading of variable 1 on the second factor has dropped considerably, it is still substantial. However, this seems to be about the best solution we can achieve so we must simply accept that 'standing height' is a hybrid variable, loading significantly both on 'leg length' and 'trunk length'. This is hardly surprising, as we mentioned previously.

After these preliminary remarks we next discuss how the technique of principal factor analysis is carried out.

4.2 A numerical example of the principal factor method: initial component analysis

A principal components analysis provides a very convenient starting point (in terms of determining suitable communality values) for the principal factor solution. Although it is not essential to start in this way, e.g. we could estimate the communalities in the same way as in Chapter 2, the following approach is very common.

To illustrate the method, we have chosen a simple example based on Blau and Duncan [1967], involving five socio-economic variables and the relationship between father and son status. The inter-correlations between these variables are shown in Table 4.4. Firstly a principal components analysis is carried out, the details of which are not shown since they are identical to those of a principal factor analysis; the results of this analysis are given in Table 4.5.

TABLE 4.4 *Correlation matrix between five socio-economic variables relating to father–son status*

	Variable				
	1	2	3	4	5
1. Father's educational level	1.00	0.52	0.45	0.33	0.33
2. Father's occupational status	0.52	1.00	0.44	0.42	0.41
3. Son's educational level	0.45	0.44	1.00	0.54	0.60
4. Son's first job status	0.33	0.42	0.54	1.00	0.54
5. Son's later job status	0.33	0.41	0.60	0.54	1.00

TABLE 4.5 *Principal components solution for five socio-economic variables*

Variable	Principal component C_1	C_2	C_3	h^2	h_2^2
1	0.68	− 0.59	− 0.31	0.91	0.81
2	0.73	− 0.41	0.45	0.90	0.71
3	0.82	0.18	− 0.34	0.82	0.70
4	0.76	0.34	0.30	0.78	0.69
5	0.77	0.40	− 0.10	0.76	0.75
% variance extracted	57	16	10		

4.3 Extraction of first factor loadings

For simplicity we will consider only the first two components in the analysis (the third accounts for barely 10 per cent of the variance). The sums of squared factor or component loadings (using the first two components) are shown in the last column of Table 4.5. These values can now be entered into the principal diagonal of the original correlation matrix in place of the unities and the method of principal factor analysis carried out. This amended correlation matrix, R, is shown in Table 4.6.

TABLE 4.6 *Correlation matrix R, between five socio-economic variables*

Variable	1	2	3	4	5
1	0.81	0.52	0.45	0.33	0.33
2	0.52	0.71	0.44	0.42	0.41
3	0.45	0.44	0.70	0.54	0.60
4	0.33	0.42	0.54	0.69	0.54
5	0.33	0.41	0.60	0.54	0.75
S_{j1}	2.44	2.50	2.73	2.52	2.63
α_{j1}	0.89	0.92	1.00	0.92	0.96

Following the notation of previous chapters, let the elements of this matrix be r_{ij} (where i refers to rows and j to columns). The initial step in the procedure for finding the first factor loadings begins by summing each column of the correlation matrix R to obtain a set of five sums S_{j1}, where the suffix j indicates the column number (i.e. the variable) and the 1 refers to the fact that we are dealing with the first factor. We then obtain a set of values α_{j1}, using the expression:

$$\alpha_{j1} = S_{j1}/S_{j1}\ (\text{max}). \tag{4.1}$$

That is, dividing each column sum by the maximum value of S_{j1} (regardless of sign). These values are given in the bottom two lines of Table 4.6.

These α_{j1} values form the basis from which the first factor loadings are calculated. However, they are only approximations to the final values.

The next step is to produce a new matrix ${}_1A$ (the prefix 1 indicating the first factor), using the values in the inter-correlation matrix R. The elements of ${}_1A$ are given by:

$${}_1a_{ij} = \sum_{k=1}^{n} r_{ik} r_{jk} \tag{4.2}$$

where, as usual, i indicates row and j column, and n is the number of variables. Thus, for example:

$$\begin{aligned} {}_1a_{23} &= r_{21}r_{31} + r_{22}r_{32} + r_{23}r_{33} + r_{24}r_{34} + r_{25}r_{35} \\ &= (0.52 \times 0.45) + (0.71 \times 0.44) + (0.44 \times 0.70) + \\ &\quad (0.42 \times 0.54) + (0.41 \times 0.60) \\ &= 1.33. \end{aligned}$$

The columns of matrix ${}_1A$ are then summed to produce a new set of values S_{j1} and equation (4.1) is used to obtain a further set of values for α_{j1}. The procedure of converting matrix R into matrix ${}_1A$ is called 'squaring the matrix'. Table 4.7 shows the matrix ${}_1A$ and the new values for S_{j1} and α_{j1}.

TABLE 4.7 *The matrix ${}_1A$ for the first factor with the new values for S_{j1} and α_{j1}*

Variable	1	2	3	4	5
1	1.35	1.26	1.28	1.13	1.18
2	1.26	1.31	1.33	1.23	1.26
3	1.28	1.33	1.54	1.41	1.49
4	1.13	1.23	1.41	1.34	1.38
5	1.18	1.26	1.49	1.38	1.49
S_{j1}	6.20	6.39	7.05	6.49	6.80
α_{j1}	0.88	0.91	1.00	0.92	0.96

It will be noticed from Table 4.7 that the matrix ${}_1A$ is symmetrical about the principal diagonal, hence only the values in and above the principal diagonal need be calculated; it will also be noted that the values for α_{j1} in Table 4.7 are different from those in Table 4.6. The matrix ${}_1A$ is now squared again to give ${}_1B$.

This procedure should be repeated as often as necessary until the successive values of α_{j1} are as close as is desired. In our case, since we are working to two places of decimals, we require that successive values of α_{j1} should be identical to two places of decimals. In this example, only one further squaring is necessary. The elements of ${}_1B$, ${}_1b_{ij}$, are given by a similar formula to (4.2), i.e.

$$ {}_1b_{ij} = \sum_{k=1}^{n} {}_1a_{ik}\,{}_1a_{jk}. $$

Adding the columns of ${}_1B$ gives us a third set of values for S_{j1} and the application of formula (4.1) gives us a third set of values for α_{j1}, which turn out to be identical to the second set. Hence we may now proceed to the next stage of the analysis, which consists of finding a set of five values, Q_{j1}, obtained from the formula:

$$ Q_{j1} = \sum_{k=1}^{n} r_{jk}\alpha_{k1}, \tag{4.3} $$

e.g.

$$ \begin{aligned} Q_{31} &= r_{31}\alpha_{11} + r_{32}\alpha_{21} + r_{33}\alpha_{31} + r_{34}\alpha_{41} + r_{35}\alpha_{51} \\ &= (0.45 \times 0.88) + (0.44 \times 0.91) + (0.70 \times 1.00) \\ &\quad + (0.54 \times 0.92) + (0.60 \times 0.96) \\ &= 2.57. \end{aligned} $$

These values are then converted into yet another set of values for α_{j1}, by means of a formula analogous to (4.1), i.e.

$$ \alpha_{j1} = Q_{j1}/Q_{j1}\ (\text{max}). $$

If this new set of values for α_{j1} differs appreciably from the previous set, then the procedure just described is repeated, i.e. the elements of R are multiplied by the new set of values for α_{j1} by means of formula (4.3) until successive sets of values for α_{j1} agree to the desired degree (in our case, to two places of decimals). In our example, the procedure only has to be carried out once. Table 4.8 gives details of the calculations.

TABLE 4.8 *Successive operations on α_{j1}*

α_{j1}	Q_{j1}	α_{j1}	F_{j1}
0.88	2.26	0.86	0.67
0.91	2.34	0.91	0.70
1.00	2.57	1.00	0.77
0.92	2.37	0.92	0.71
0.96	2.48	0.96	0.74

The final column of Table 4.8, F_{j1}, gives the values for the first factor loadings.

These are obtained from the formula:

$$F_{j1} = \alpha_{j1}\sqrt{\lambda_1 / \sum_j \alpha_{j1}^2}, \tag{4.4}$$

where λ_1 is the value of Q_{j1} corresponding to that value of α_{j1} which equals unity. In our example, therefore, $\lambda_1 = 2.57$.

4.4 Expected values after computation of first factor

In the centroid method discussed in Chapter 2, equation (2.1) (re-stated here for convenience),

$$r_{ij} = \sum_k F_{ik} F_{jk}$$

relates the original correlations between variables, r_{ij}, to the correlations between the variables and factors. In a similar way, we may in the principal factor analysis proceed to test the hypothesis that only one factor is needed to account for the original r_{ij}'s. As with the centroid method, a matrix of expected values is set up, the elements of which are given by;

$$_1e_{ij} = F_{i1} F_{j1}.$$

The first matrix of expected values E_1 is shown in Table 4.9.

TABLE 4.9 *First expected values matrix E_1*

Variable	1	2	3	4	5
1	0.45	0.47	0.52	0.48	0.50
2		0.49	0.54	0.50	0.52
3			0.59	0.55	0.57
4				0.50	0.53
5					0.55

The next step, exactly as in the centroid method, is to determine the first residual matrix, R_2, whose elements $_1s_{ij}$ are given by:

$$_1s_{ij} = r_{ij} - {_1e_{ij}}.$$

We again sum the columns to find S_{j2} and hence determine α_{j2} as before (with the second factor, $\alpha_{j2} = S_{j2}/S_{j2}$ (max.)). The first residual matrix is given in Table 4.10.

TABLE 4.10 *First residual matrix, R_2*

Variable	1	2	3	4	5
1	0.36	0.05	−0.07	−0.15	−0.17
2	0.05	0.22	−0.10	−0.08	−0.11
3	−0.07	−0.10	0.11	−0.01	0.03
4	−0.15	−0.08	−0.01	0.19	0.01
5	−0.17	−0.11	0.03	0.01	0.20
S_{j2}	−0.08	−0.12	−0.04	−0.04	−0.04
α_{j2}	0.67	1.00	0.33	0.33	0.33

4.5 Extraction of second factor

Now we repeat the procedure given for the extraction of the first factor loadings, i.e. squaring the matrix R_2, obtaining values of Q_{j2} etc., until a stable set of values of α_{j2} emerges, which can then be used to obtain the second factor loadings. Here we run into a slight difficulty: squaring the matrix R_2 to produce $_2A$ produces values which are small and therefore rounding errors (since we are working to only two places of decimals) become large relative to them. Normally, one would work to four or more places of decimals to obviate this difficulty, but in any case it does not matter since the consequence is that the process of finding a stable set of values for α_{j2} merely takes a little longer.

Since quite a large number of these iterative calculations are required, we show only the first and final stages in the process in Table 4.11. The values of Q_{j2} are obtained from an expression similar to (4.3), i.e.

$$Q_{j2} = \sum_k {}_1s_{jk}\,\alpha_{k2}.$$

TABLE 4.11 *Successive operations on α_{j2}*

α_{j2}	Q_{j2}	α_{j2}		α_{j2}	Q_{j2}	α_{j2}	F_{j2}
0.67	0.16	1.00	...	1.00	0.59	1.00	0.55
1.00	0.16	1.00	...	0.51	0.30	0.51	0.28
0.33	−0.10	−0.63	...	−0.27	−0.16	−0.27	−0.15
0.33	−0.12	−0.75	...	−0.49	−0.29	−0.49	−0.27
0.33	−0.14	−0.88	...	−0.63	−0.37	−0.63	−0.35

The second factor loadings, shown in the last column of Table 4.11, are calculated from an expression similar to (4.4), i.e.

$$F_{j2} = \alpha_{j2}\sqrt{\lambda_2/\sum_j \alpha_{j2}^2},$$

where $\lambda_2 = 0.59$, i.e. the value of Q_{j2} corresponding to the highest value of α_{j2} (1.00).

As before, we can test the hypothesis that the first residual correlations can be accounted for by the correlations between the variables and the second factor. Hence a matrix of second expected values ${}_2E$, with elements ${}_2e_{ij}$ (where the prefix 2 refers to the second factor), is obtained from the expression;

$${}_2e_{ij} = F_{i2}F_{j2},$$

and a second residual matrix R_3, with elements ${}_2s_{ij}$, is calculated using the expression:

$${}_2s_{ij} = {}_1s_{ij} - {}_2e_{ij}.$$

Both matrices are shown in Table 4.12.

TABLE 4.12 *Second expected value and residual matrices*

	${}_2E$, second expected value matrix					R_3, second residual matrix				
Variable	1	2	3	4	5	1	2	3	4	5
1	0.30	0.15	−0.08	−0.15	−0.19	0.06	−0.10	0.01	0.00	0.02
2		0.08	−0.04	−0.08	−0.10		0.14	−0.06	0.00	−0.01
3			0.02	0.04	0.05			0.09	−0.05	−0.02
4				0.07	0.09				0.12	−0.08
5					0.12					0.08

Although the residuals are not quite as small as one would like, it is advisable to re-estimate the communalities, rather than to extract a third factor, which will certainly not account for greater than 10 per cent of the variance. The factor matrix and resulting communalities appear in Table 4.13.

If we compare Table 4.5, the principal component analysis, and Table 4.13, it will be seen that the communalities differ appreciably.

The principal factor analysis may be repeated using the set of communality estimates obtained from the first solution (Table 4.13). This procedure (of iteration) can be repeated as often as necessary, until the communalities arising from successive analysis no longer differ appreciably. In our example, this is found to happen after seven iterations have been carried out. Table 4.14 gives the final set of factor loadings and communalities.

TABLE 4.13 *Principal factor solution for father–son status example: first iteration*

Variable	F_1	F_2	h_2
1	*0.67*	*0.55*	0.75
2	0.70	0.28	0.57
3	*0.77*	− 0.15	0.62
4	*0.71*	− 0.27	0.58
5	*0.74*	− *0.35*	0.67
% variance extracted	51.4	12.0	

TABLE 4.14 *Principal factor solution after seven iterations*

Variable	F_1	F_2	h_2
1	*0.67*	*0.53*	0.71
2	*0.63*	0.17	0.43
3	*0.76*	− 0.13	0.59
4	*0.67*	− 0.21	0.50
5	*0.72*	− *0.31*	0.62

The final residual matrices for the principal factor solution after seven iterations, for the principal components solution (first two components only) and for the principal components solution (first three components only) are shown in Tables 4.15(*a*), (*b*) and (*c*) respectively.

TABLE 4.15 *Final residual matrices*

Variable	1	2	3	4	5
1	0.00	0.00	0.01	− 0.01	0.00
2		0.00	− 0.02	0.02	0.00
3			0.00	0.00	0.01
4				0.00	− 0.01
5					0.00

(*a*)

Variable	1	2	3	4	5
1	0.19	− 0.23	0.00	0.02	0.03
2		0.29	− 0.09	0.00	0.01
3			0.30	− 0.14	− 0.01
4				0.31	− 0.18
5					0.25

(*b*)

Variable	1	2	3	4	5
1	0.09	− 0.09	− 0.11	0.07	0.00
2		0.09	0.06	− 0.14	0.04
3			0.18	− 0.04	− 0.13
4				0.22	− 0.15
5					0.24

(*c*)

4.6 Comparison between principal components and principal factor solutions to the socio-economic problem on father - son status

Table 4.15 gives the residuals following the extraction of the first two components (4.15(c)) and first three components (3.15(c)) of the original principal component analysis. It can be seen that the principal factor analysis has accounted for the original correlations between the five variables much more satisfactorily than the principal component analysis, and also more parsimoniously, since only two factors are required rather than three components. Thus the somewhat different aims of the two types of analysis (to account for the inter-correlations parsimoniously, as opposed to accounting for the total variance of each variable) can lead to different results and hence different interpretations. In general, the principal factor analysis leads to more parsimonious solutions, this being more likely when there is a relatively high degree of unique variance associated with each variable (when communalities are low). It should also be clear that the choice of communalities can affect the outcome of a factor analysis. For this reason, it is advisable to include as large a number of variables as possible, with a fair sprinkling of high correlations between variables. This ensures that a large proportion of the variance of each variable will be common variance (communalities will be high) and therefore there is less chance of obtaining different solutions depending upon which estimate of communalities one chooses. In the body-size problem, the principal components solution was essentially similar to the centroid solution because communalities were rather high.

4.7 Interpretation of factors

From Table 4.14, F_1 emerges as a general factor with high loadings for each variable; this may be called 'family status'. The second factor is bipolar, with father's educational level at one pole and son's later job level at the other. This is interesting and suggests the possible existence of an underlying variable tending to act on these observed variables in opposite directions. However, it is very difficult to interpret this factor as it stands and, following the arguments concerning rotation in the previous chapter, we will proceed to rotate the factors in the hope that a clearer picture will emerge.

4.8 Rotation of factors

As with previous examples, rotation may aid the task of interpretation. The results of the present example were in fact used in the discussion of the rotation problem in the previous chapter. There it was mentioned that an orthogonal rotation did not enable us to achieve simple structure whilst an oblique rotation did. The oblique factor loadings are reproduced in Table 4.16.

TABLE 4.16 *Oblique factor loadings: father – son status*

Variable	F_1'	F_2'
1	*0.81*	– 0.06
2	*0.48*	– 0.29
3	0.29	– *0.60*
4	0.18	– *0.60*
5	0.12	– *0.71*

Oblique rotation has caused two clusters of variables to emerge rather clearly. The not very exciting conclusion (in this instance) appears to be that F_1 may be called 'father status', and F_2 may be called 'son's status'. In other words, variables pertaining to the father are more closely related to each other than they are to those relating to the son, and vice versa. Moreover, the two factors are correlated to some extent, showing that some underlying variable (second-order factor) determines both father's and son's status. The nature of this influence is such that when the father's status is high, the son's also tends to be high. We can see that this is the case by referring back to Fig. 3.2. The dotted axes A and B pass through the centroids of the two clusters we have isolated. Since the angle between A and B is acute, it is clear that the two axes are positively correlated, i.e. that the two clusters are fairly close together and therefore possess a certain similarity. If the angle between A and B had been obtuse, then the two clusters would have been widely separated in space and negatively correlated. Had this been the case, the interpretation would have been that as the father's status increases, the son's decreases. If A and B had been approximately orthogonal, then it is likely that an orthogonal rotation would have sufficed to produce simple structure and no oblique rotation would therefore have been necessary.

4.9 Other methods of factor analysis

Two methods of factor analysis have been presented so far (not including the related principal components method) and their similarities and differences described. The principal factor analysis method is probably the most widely used today, since computers have taken over the task of carrying out the formidable amount of computational labour involved. Both the principal factor and centroid methods require estimates of communalities to be made.

Another approach is to estimate the number of common factors at the start, rather than the communalities. One such method is the maximum-likelihood solution devised originally by Lawley [1940] which produces the communalities as a by-product, and a test of significance is available which enables one to test whether the number of common factors assumed initially was sufficient. Another method of this type, described in detail in Harman [1967], is called the 'minres' solution. This seeks to minimise the sum of squares of residuals, and hence to reproduce accurately the original correlations; again, communalities emerge as a by-product of the analysis. Neither of these methods can be discussed here, and the interested reader is referred to more advanced texts.

Another method, described in Comrey [1973], and known as the 'minimum residual method', utilises only off-diagonal entries in the correlation matrix so that no communality estimates are required. In addition, it is not necessary to estimate the number of factors to be extracted. As with the preceding methods, communality estimates emerge as a by-product of the analysis. This method is similar to the principal components and principal factor methods except that the entries in the principal diagonal differ: for principal components, the entries are unity; for principal factors they are communality estimates; and for minimum residuals, they are zeros. With small matrices, markedly different results may be obtained by the three methods because the ratio of the number of diagonal entries to that of other entries is large (and their importance is correspondingly large), whilst for large matrices the results yielded by the three methods may be very similar, since the above ratio is smaller.

Yet another approach is the multiple-group solution, an example of which will be given in the next chapter. All the former methods described above serve as initial solutions which are then rotated to identify 'clusters' of closely-related variables. The multiple-group solution locates these clusters directly, and the method requires an estimate of number of factors and of the communalities.

The foregoing represent the most widely used methods of factor analysis. Many other varieties exist but they are mainly of historic interest since the advent of high-speed computers. We will mention one more which is interesting because of its great simplicity in computation; this is the so-called square-root method.

TABLE 4.17 *Inter-correlations between five socio-economic variables in the father–son status example*

Variable	1	5	4	3	2
1	0.73	0.32	0.33	0.45	0.52
5		0.62	0.54	0.60	0.41
4			0.50	0.54	0.42
3				0.59	0.44
2					0.43

Using the example of the five socio-economic variables, the correlation matrix of Table 4.6 is re-arranged (for a reason to be explained later), and the communalities estimated. The communality estimates arising from the seventh iteration of the principal factor analysis (Table 4.14) have been employed in the present example. The initial correlation matrix appears in Table 4.17.

4.10 The square-root method: extraction of first factor loadings

We first find the square root of the communality of variable 1, which we label h_1. Thus,

$$h_1 = \sqrt{0.73} = 0.85, \text{ in this case.}$$

Then the first factor loadings are given simply by

$$F_{1j} = r_{1j}/h_1, \tag{4.5}$$

where the suffix j denotes the variable number. Table 4.18 gives the complete set of first factor loadings.

TABLE 4.18 *First factor loadings*

	Variable (j)					
	1	5	4	3	2	% variance extracted
F_{1j}	0.85	0.38	0.39	0.53	0.61	33

What in effect this procedure achieves is to explain *all* the common variance of variable 1 in terms of one factor. Since variable 1 therefore has zero factor loadings on any other factor, all of the correlations between it and the other variables are explicable in terms of their factor loadings on F_1 (by equation (2.1)). The first row of the first residual matrix will therefore consist entirely of zeros. A table of expected values (on the assumption that only one factor is necessary to explain the correlations) is constructed in the usual way, i.e. using:

$$_1e_{ij} = F_{1i}F_{1j}.$$

This appears in Table 4.19. Then a table of first residuals is constructed using:

$$_1s_{ij} = r_{ij} - {_1e_{ij}}.$$

This appears in Table 4.20.

TABLE 4.19 *First expected values E_1*

Variable	1	5	4	3	2
1	0.73	0.32	0.33	0.45	0.52
5		0.14	0.15	0.20	0.23
4			0.15	0.21	0.24
3				0.28	0.32
2					0.37

TABLE 4.20 *First residuals R_1*

Variable	1	5	4	3	2
1	0.00	0.00	0.00	0.00	0.00
5		0.48	0.39	0.40	0.18
4			0.35	0.33	0.18
3				0.31	0.12
2					0.06

4.11 Extraction of the second factor loadings

The square root of the communality of the variable in the second row is now calculated, $h_5 = 0.69$. The second factor loadings are now found from:

$$F_{2j} = {}_1s_{5j}/\sqrt{{}_1s_{55}}. \tag{4.6}$$

Table 4.21 gives the second factor loadings.

TABLE 4.21 *Second factor loadings F_{2j}*

	Variable (j)					
	1	5	4	3	2	% variance extracted
F_{2j}	0.00	0.69	0.57	0.58	0.26	24

This procedure has the effect of explaining all the residual common variance of variable 5 in terms of its loading on the second factor. Since variable 5 loads only on F_1 and F_2 and has zero loadings on any subsequent factors, it follows that all the residual correlations which is has with the other variables (4, 3 and 2) are explicable in terms of their second factor loadings, and that the second row of the second residual matrix will consist of zeros. Matrices of second expected values and second residuals may be constructed in the normal way, i.e. using the formulae:

$$_2e_{ij} = F_{2i}F_{2j}$$

and

$$_2s_{ij} = {_1s_{ij}} - {_2e_{ij}},$$

respectively. These matrices appear in Table 4.22.

TABLE 4.22 *Second expected values E_2, and second residuals R_2*

Variable	E_2				
	1	5	4	3	2
1	0.00	0.00	0.00	0.00	0.00
5		0.48	0.39	0.40	0.18
4			0.32	0.33	0.15
3				0.34	0.15
2					0.07

Variable	R_2				
	1	5	4	3	2
1	0.00	0.00	0.00	0.00	0.00
5		0.00	0.00	0.00	0.00
4			0.03	0.00	0.03
3				− 0.03	− 0.03
2					− 0.01

Since the second residuals are so small, we may conclude that only two factors are required to explain the correlations. In order to show the similarity between the square-root solution and the principal-factor solution, an oblique rotation has been carried out on the square-root factors. These are presented along with the oblique rotation of the principal factor solution (from Table 4.16) in Table 4.23.

It can be seen from this that the correspondence is very close. This will not always be the case since the square-root method does not ensure that the maximum variance is extracted by each successive factor, and this sometimes means

TABLE 4.23 *Comparison of oblique square-root and principal factors*

Variable	Square-root		Principal factor	
	F_1	F_2	F_1	F_2
1	*0.79*	− 0.04	*0.81*	− 0.06
2	*0.47*	− 0.29	*0.48*	− 0.29
3	0.28	− *0.61*	0.29	− *0.60*
4	0.16	− *0.59*	0.18	− *0.60*
5	0.11	− *0.71*	0.12	− *0.71*
Angle between factors	72°		72°	
correlation between factors	0.31		0.31	

that more factors will need to be extracted before the residuals are reduced to near zero. Where it scores is in the ease of computation, but with the advent of computers this is no longer a problem.

The reader may be puzzled about the re-arrangements of variables in the original correlation matrix. The reason for this is that if the variables are not ordered carefully, one can find oneself with residual elements in the principal diagonal which are negative, in which case one cannot calculate their square roots. To guard against this, one needs to ensure that, following the extraction of a factor, the diagonal element in the subsequent residual matrix, that is to be used in the extraction of the next factor, is left as large as possible. Variable 1 was selected in the first row of Table 4.17 because its communality was high whilst its correlations with the other variables were fairly low, and the order of the variables in columns 2 - 5 was made 5, 4, 3 and 2 because this was in the increasing order of magnitude of their correlations with variable 1. The combination of a high value for h_1 and a low value for r_{15} ensures that F_{15} will be low (from equation (4.5)), and therefore the value to be employed in finding the second factor loadings, S_{55} (see equation (4.6)) will be high.

4.12 Uniqueness of solutions

It is clear that the factor analysis of inter-correlations between variables does not produce unique solutions. Different investigators can arrive at different conclusions because of differences in methods employed, communality estimates, or the position of rotated axes. The particular variables chosen for study can affect the outcome, i.e. adding or subtracting variables can affect the pattern of factors which emerge. A number of points may be considered which will help with this problem of non-uniqueness:

(1) It is advisable in any factor analytic study to have a large number of observations on which each correlation is based. This makes it more likely that the correlation coefficients are similar to their population values.

(2) One should try to ensure that the correlation coefficients are stable, i.e. the measurements on which they are based are reliable. This decreases the amount of unique variance (which may be considered to include error variance) associated with each variable.

(3) One should include as many variables as possible; the more there are, the greater the common variance will be. This makes it more likely that all important factors will emerge from the analysis.

(4) It is worth employing several methods to see whether roughly equivalent solutions emerge.

4.13 Geometrical interpretation of principal factor analysis

It is beyond the scope of this book to derive any of the expressions used in the factor analysis described. However, it may be helpful to describe briefly what is achieved by the algebraic manipulations in geometric terms. As was mentioned in Chapter 3, variables can be regarded as points in space, the axes of which are the factors. The centroid method ensures that the first factor passes through the centroid of these points, and that subsequent factors are orthogonal to the first and to each other. The principal factor and principal component solutions ensure that the first factor passes through the points in such a way that the perpendicular distances from the points to the factors are made as small as possible or, more precisely, that the sum of squared distances from points to the axis is a minimum; this ensures that the first factor extracts or explains the maximum variance. As with the centroid method, subsequent factors are orthogonal to the first and to each other.

Suppose that we have a set of variables (points) which form a scatter of points in three-dimensional space (i.e. only three factors are required to explain the correlations between them) and that the scatter is roughly ellipsoid in shape; that is to say, it looks rather like a rugby football. The axes representing the factors arising from a principal factor (or principal components) solution would then be as follows: the first axis (extracting the most variance) would pass through the two narrow ends of the rugby ball, i.e. be the major axis of the ellipsoid, whilst the other two axes would pass through a plane parallel with the middle cross-section of the ball at a right angle to each other, i.e. be the minor axes of the ellipsoid.

4.14 Some practical examples of the techniques

It may be useful to conclude with some actual examples taken from literature. We start with work done by Giggs [1973] in the field of social geography. In this study, 'The spatial variations in the distribution of schizophrenia within Nottingham are identified. Principal factor analysis is used to relate these patterns to aspects of the man-made environment and to relevant demographic and socio-economic phenomena.' The author stresses the advantages of oblique rotation to simple structure rather than orthogonal rotation using, for example, the Varimax procedure.

The analysis was based on a sample of 444 persons classed as schizophrenics who were sent to hospital for the first time from an address within the city boundary. In all, 41 variables were employed, which may be divided into six groupings. The first group consisting of 12 variables related to all first admissions, sex, age, marital status, and place of birth; chronic cases were also identified. The remaining 29 variables were socio-environmental. The second group related to population structure (distribution by age, sex, marital status, whether foreign born, colour, etc.); the third group consisted of four socio-economic attributes (social class, whether unemployed, etc.); the fourth group of four variables related to type of household tenure (owner-occupied, rented, etc.); the fifth group of nine variables related to household and housing characteristics (sharing facilites, outside w.c., density of occupation, etc.); while the last group contained one variable measuring distance from city centre. Detailed results of the analysis for primary orthogonal, orthogonal rotated (Varimax) and for oblique rotation using Promax are presented in the paper. The first orthogonal factor is, as expected, too general in character to be usefully interpreted. However, some improvement is achieved with Varimax orthogonal rotation; only 20 of the variables have significant loadings on the first factor, with the highest being related to schizophrenia; this first factor may thus be interpreted as a 'schizophrenic' dimension. Further improvement in this first factor is achieved with oblique rotation using the Promax procedure; only 14 of the variables have significant loadings. Five further factors are extracted using Promax. The second is identified as a 'stage of life' factor, i.e. it distinguishes middle- and old-aged persons with small households, from young persons, large households, the 'mobile-married', etc. The third factor is labelled 'urbanism-familism' and distinguishes rooming-house areas, with small dwellings, single-mobile persons, the divorced, coloured immigrants, etc., from large dwellings, large households and married persons. The remaining factors are labelled 'rented housing – housing amenities' and 'socio-economic', whilst the sixth factor is considered unimportant.

All the factors are related spatially to the geography of Nottingham and the subsequent spatial distributions examined.

The author then goes on to determine second-order factors. He concludes that most of the first-order factors can be regarded as 'parts of a single second-order factor', accounting for 86.8 per cent of the common variance. The author then examines the loadings of the first-order factors on this second-order 'major dimension' and states that the second-order Factor I can be termed a 'social/environmental-schizophrenia' dimension.

The statistical and cartographic evidence, the author summarises, shows that 'the majority of both first admission and chronic schizophrenic patients entered hospital from an area characterised by high levels of urbanism and rented housing'.

The study is interesting in two ways. Firstly, because it discusses and illustrates the advantages, in the author's opinion, of oblique as opposed to orthogonal rotation, and secondly because it shows how second-order factors can be interpreted in terms of first-order factors.

The second example is in the use of principal factor analysis and is due to Ziegler and Atkinson [1973], in a study of political liberal and conservative attitudes. The authors examine the connection between political attitudes and the level of information of the subject. In particular they are concerned with examining the theory of Kerlinger that liberalism and conservatism are two unipolar orthogonal factors, rather than a single bipolar dimension. The authors' two hypotheses are firstly that persons better informed about political affairs will, as a group, have higher correlations among different aspects of political-economic liberalism-conservatism; and secondly that second-order factor analysis for high and low information groups will yield only one factor, interpretable as a liberalism-conservatism; Ziegler used a 113-item Conservatism Scale which measured 17 relatively distinct aspects of liberalism-conservatism (for example, individual rights versus government control; fear of change; unionism; classical versus Keynesian economics; etc.). Further, each subject was asked to assess their opinions, on a seven-point scale, on such matters as international issues, domestic issues, etc. They were also asked their attitude to 17 laws considered by Congress, each law corresponding to one of the sub-scales in the Conservatism Scale.

Finally, two other relevant measures were used, a 'level of information' scale of 28 factual items of domestic political affairs and an 'interest' scale of three items, used as a check on the notion that information and interest are related. The subjects chosen were 369 university students, 40 per cent second year, 30 per cent first year, and 30 per cent final year. These students were divided, on the basis of the information test, into highly informed and 'lowly' informed.

We are concerned here mainly with the second hypothesis, which is tested using principal factor analysis. (The first hypothesis is tested using comparatively straightforward correlation analysis and found to be supported.) The author divided his sample into the two separate groups (high information and low information) and performed a principal factor analysis on each group separately, together with oblique rotations of those components with latent roots greater than unity.

For the high information group, four factors were identified. The first is a 'general liberalism' factor; the second can be labelled as 'limiting central government power or decentralization' and is predominantly conservative. The third factor, also conservative, is concerned with support for the military-police complex; the fourth is difficult to interpret, but seemed to be conservative and concerned with preserving the *status quo*.

For the low information group, five factors were identified. The first was a general conservatism factor, the second a general liberalism factor; the third factor is conservative and concerned with decentralisation; factor four is also conservative and is related to items opposing dissent; the fifth factor has only one variable with a significant loading referring to ease of amending the constitution. Of particular interest is the factor correlation matrix, with the low correlation between the liberalism factor and the conservatism factor for the high information group and the low correlation between the first two factors for the low information group. That is, the liberal and conservatism items do appear to lie on separated, orthogonal dimensions, supporting Kerlinger.

However, the important analysis is the second-order one, achieved by factor analysing the oblique rotated factor correlation matrix. For the high information group two independent second-order oblique factors were identified, the first predominantly conservatism, the second a bipolar liberalism-conservatism factor; these factors are correlated. Two second-order oblique factors were identified for the low information group, the first essentially a conservatism factor, the second factor somewhat mixed but predominantly liberalism; these factors are independent.

Thus, the hypotheses that second-order analysis for the two information levels will give one primary component is not supported. The author draws three main conclusions. Firstly, among a population of university students, political attitudes are multidimensional. Secondly, in particular, liberalism and conservatism tend to be independent. Thirdly, better informed students' political attitudes are more constrained and tend towards bipolarity. Kerlinger's hypothesis seems to hold best for relatively uninformed persons.

The important lesson for the reader is that different subgroups within the general population have different attitudes, and attempts at analysis which ignore these differences and treat the population under study as a homogenous set will yield unreliable results.

Finally, we refer briefly to a principal components analysis by Ahamad [1967] in the field of criminology. The author's purpose is to 'investigate the relationships between several different crimes and to determine to what extent the numbers of crimes from year to year may be explained by a small number of unrelated factors'. The study covered the incidence of 18 groups of offences over a period of 14 years. The first component extracted is interpreted as a 'population increase' component, in particular for the 13 - 19 years age group. The second component is more difficult to identify; the author concludes that this component is an error-component, reflecting changes in methods of recording offences. Finally, the author extracts a third component (the three components together accounting for 92 per cent of the total variance), but he finds it impossible to interpret this in any meaningful way. He concludes his investigation by saying that population growth is the factor which accounts for recent increases in crime.

5 Multiple Groups Analysis

5.1 Introduction

The previous chapters have dealt with methods from which orthogonal factors emerge. Clusters of related variables are then obtained subsequently by orthogonal or oblique rotations to simple structure. This chapter is concerned with a method called *multiple groups analysis* from which an oblique factor matrix is obtained directly and all factors are extracted simultaneously. The number of such groups (and therefore factors) is assumed at the outset either on the basis of *a priori* knowledge of the field of study or by a systematic procedure which the interested reader can find in more advanced texts, such as Harman [1967]. It is, however, not serious if an incorrect assumption about the number of groups is made initially.

Furthermore, the concept and meaning of *factor patterns* will be introduced which leads to an alternative method for identifying clusters other than that of rotation to simple structure. Finally, another simpler method of locating clusters of related variables, known as *linkage analysis*, will be described.

To illustrate the method of multiple groups analysis we consider a matrix of correlations between eight tests of mental ability taken from Child [1970] and shown in Table 5.1. In this table, communality estimates are simply the highest correlation in each row of the completed matrix.

Tests 1 and 2 are conventional 'intelligence tests' purporting to measure convergent thinking. The others are tests of fluency (F) and originality (O) in divergent thinking. Tests 3 - 6 require verbal responses: whilst tests 7 and 8 require non-verbal responses. Suppose, on the basis of psychological theory, we believe that these tests measure three relatively distinct aspects of mental ability, i.e. convergent thinking, verbal divergent thinking and non-verbal divergent thinking. That is to say, we believe that there are three factors and that they indicate three clusters of variables (1, 2; 3, 4, 5, 6; and 7, 8). We can proceed to test this hypothesis by means of the multiple-group method. In a sense, this method is closer to the spirit of scientific research than previous methods because we are starting from a definite theory about the nature of the variables under study and testing it. Here is yet another use for factor analysis, besides being a tool for giving parsimonious descriptions of data and as a starting-point for theory construction.

TABLE 5.1 *Correlations between eight tests of mental ability*

	1	2	3	4	5	6	7	8
1. AH5 - verbal	0.54	0.54	0.08	0.18	0.20	0.13	0.10	0.05
2. AH5 - spatial	0.54	0.54	0.01	0.05	0.07	−0.01	0.08	0.00
3. Uses - F	0.08	0.01	0.58	0.58	0.51	0.26	0.46	0.22
4. Uses - O	0.18	0.05	0.58	0.58	0.46	0.40	0.27	0.22
5. Consequences - F	0.20	0.07	0.51	0.46	0.51	0.46	0.40	0.21
6. Consequences - O	0.13	−0.01	0.26	0.40	0.46	0.46	0.11	0.18
7. Circles - F	0.10	0.08	0.46	0.27	0.40	0.11	0.51	0.51
8. Circles - O	0.05	0.00	0.22	0.22	0.21	0.18	0.51	0.51

5.2 Producing the oblique factor matrix

Firstly, we divide the variables into the three groups corresponding to our hypothesis. Taking each variable in turn, we sum the correlations which it has with all the variables in the first group, then those which it has with the second group, and finally those which it has with the third group; these sums include the communalities. Let these sums be represented by S_{j1}, S_{j2}, and S_{j3} respectively, for variable j. Thus, for variable 1 we have,

$$S_{11} = h_1^2 + r_{12} = 0.54 + 0.54 = 1.08$$

$$S_{12} = r_{13} + r_{14} + r_{15} + r_{16} = 0.08 + 0.18 + 0.20 + 0.13 = 0.59$$

$$S_{13} = r_{17} + r_{18} = 0.10 + 0.05 = 0.15.$$

The complete set of these sums for each variable are shown in Table 5.2.

TABLE 5.2 *Sums of correlations for three groups*

Group r	Variable j	Sum S_{jr}		
		S_{j1}	S_{j2}	S_{j3}
1	1	1.08	0.59	0.15
	2	1.08	0.12	0.08
2	3	0.09	1.93	0.68
	4	0.23	2.02	0.49
	5	0.27	1.94	0.61
	6	0.12	1.58	0.29
3	7	0.18	1.24	1.02
	8	0.05	0.83	1.02

Next we sum the entries in Table 5.2 for each column and each group separately to produce a matrix S_{rk}, where r indicates the number of the group and k the column number. For example:

$$S_{23} = 0.68 + 0.49 + 0.61 + 0.29 = 2.07.$$

The complete S_{rk} matrix is shown in Table 5.3. The final row of the table, $\sqrt{S_{kk}}$, are the square roots of the diagonal entries.

TABLE 5.3 *Sums of sums of correlations for three groups*

r \ k	1	2	3
1	2.16	0.71	0.23
2	0.71	7.47	2.07
3	0.23	2.07	2.04
$\sqrt{S_{kk}}$	1.47	2.73	1.43

Now the oblique factor loadings are obtained by simple application of the expression

$$F_{jk} = S_{jr}/\sqrt{S_{kk}},$$

where F_{jk} is the factor loading of variable j on factor k. For example, for variable 5 and factor 2, we have:

$$F_{52} = S_{52}/\sqrt{S_{22}} = 1.94/2.73 = 0.71.$$

The matrix B of factor loadings appears in Table 5.4.

TABLE 5.4 *Oblique factor loading matrix B (significant loadings in italic)*

Variable	Oblique factor		
	B_1	B_2	B_3
1	*0.73*	0.22	0.10
2	*0.73*	0.04	0.06
3	0.06	*0.71*	*0.48*
4	0.16	*0.74*	*0.34*
5	0.18	*0.71*	*0.43*
6	0.08	*0.58*	0.20
7	0.12	*0.45*	*0.71*
8	0.03	*0.30*	*0.71*

What this procedure does is to position the three axes such that they pass through the centroids of the three clusters of variables. Normally, this will mean that the axes are oblique, but this is not inevitable. Positioning the axes in this way does not help interpretation since it means that variables will load significantly on several factors, i.e. simple structure is not approximated. One could rotate the factors to a more advantageous position, but the following section shows another way of looking at the problem, by introducing a fresh concept.

5.3 Production of oblique factor pattern

B is known as the *factor structure*, and the factor loadings represent correlations between variables and factors. Another way of expressing the relationship between variables and factors is in terms of the co-ordinates of the former with respect to the latter. Figure 5.1 shows the difference.

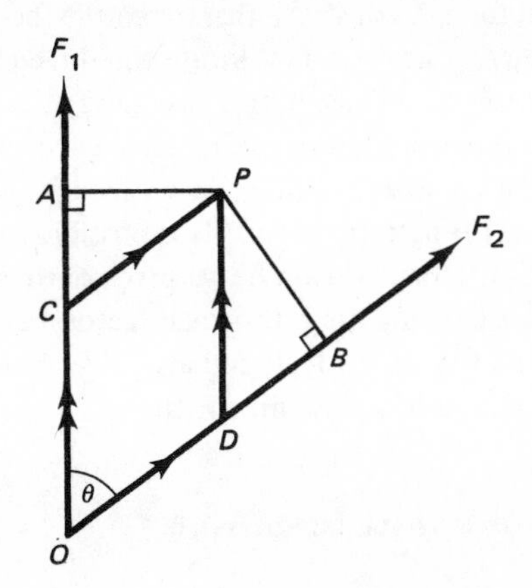

Fig. 5.1 *Factor loadings and co-ordinates of point P on two factors or axes*

Factor loadings are given by dropping perpendiculars on to the two axes. Thus, OB and OA represent the factor loadings of point (or variable) P on F_2 and F_1 respectively. The co-ordinates of P are obtained by dropping lines on to the axes which are parallel to the two axes. Thus OD and OC give the co-ordinates of P with respect to F_2 and F_1 respectively. As θ increases, the lines PC and PD get more nearly like PA and PB. When the axes are orthogonal, they coincide and the co-ordinates are then the same as the factor loadings.

How does this help? Clearly, if P forms part of a cluster through which F_2 passes, and P lies near to the axis F_2, then its F_2 co-ordinate will be high, whilst its F_1 co-ordinate will be low (although, if F_1 and F_2 correlate fairly highly, i.e. if θ is small, then the loadings of P would be high on both factors).

A matrix of the co-ordinates of the variables is called the *factor pattern*, denoted P. The factor pattern for the present example is shown in Table 5.5.

TABLE 5.5 *Factor pattern matrix, P (high values in italic)*

Variable	Factor pattern, P		
	P_1	P_2	P_3
1	*0.71*	0.11	– 0.04
2	*0.75*	– 0.12	0.05
3	– 0.07	*0.65*	0.14
4	0.03	*0.78*	– 0.08
5	0.06	*0.67*	0.07
6	– 0.02	*0.67*	– 0.15
7	0.04	0.10	*0.65*
8	– 0.03	– 0.10	*0.77*

It is clearer from the factor pattern above than from the factor structure of Table 5.4, that the three factors do in fact locate the three clusters which we originally hypothesised to exist. Thus, in this case (and in general if the *a priori* hypothesis about the number of factors is correct) the factor pattern is a distinct improvement over the factor structure in terms of distinguishing variable clusters. How do we derive the factor pattern matrix? It is possible to derive it directly from the oblique factor structure. However, we propose to employ a two - stage procedure which involves deriving the orthogonal factor matrix, since we can then use this to determine a matrix of residuals and also determine how much of the common variance is attributable to the three factors.

5.4 Production of the orthogonal factor matrix

The first step is to derive the matrix of correlations between oblique factors, which we denote Φ. The elements ϕ of this matrix are derived from Table 5.3 using the expression:

$$\phi F_r F_k = S_{rk} / \sqrt{S_{rr} \times S_{kk}}.$$

For example,

$$\phi_{23} = S_{23} / \sqrt{S_{22} \times S_{33}} = 2.07 / \sqrt{7.47 \times 2.04} = 0.53.$$

This gives the following matrix,

$$\Phi = \begin{array}{c} \\ B_1 \\ B_2 \\ B_3 \end{array} \begin{array}{c} \begin{array}{ccc} B_1 & B_2 & B_3 \end{array} \\ \begin{bmatrix} 1.00 & 0.18 & 0.11 \\ 0.18 & 1.00 & 0.53 \\ 0.11 & 0.53 & 1.00 \end{bmatrix} \end{array} .$$

From this matrix it can be seen that whereas the first factor is relatively uncorrelated with the other two (i.e. the convergent and divergent thinking tests are relatively independent), factors two and three are substantially correlated. This is hardly surprising since all the variables with high loadings on F_2 and F_3 are tests of divergent thinking.

The second stage is to produce a 'transformation' matrix T, which has the same number of rows and columns as Φ and which has zeros below the principal diagonal. Matrices of this shape are called upper triangular matrices. It has the following form:

$$T = \begin{bmatrix} t_{11} & t_{12} & t_{13} \\ 0 & t_{22} & t_{23} \\ 0 & 0 & t_{33} \end{bmatrix}.$$

The elements of T can be derived from Φ by using expressions of the following form:

$$\phi_{jj} = \sum_{i=1}^{j} t_{ii}^2 \text{ and } \phi_{jk} = \sum_{i=1}^{j} t_{ij} \times t_{ik} \text{ where } k > j \tag{5.1}$$

Actually, the first row of T is always the same as the first row of Φ, so that formulae (5.1) are only applied for $j \geqslant 2$, although (5.1) holds also for $j = 1$. Thus, for our example, we have:

$$\phi_{22} = t_{12}^2 + t_{22}^2$$

$$\phi_{23} = t_{12}\, t_{13} + t_{22}\, t_{23}$$

$$\phi_{33} = t_{13}^2 + t_{23}^2 + t_{33}^2.$$

Applying this procedure gives:

$$T = \begin{bmatrix} 1.00 & 0.18 & 0.11 \\ 0.00 & 0.98 & 0.52 \\ 0.00 & 0.00 & 0.85 \end{bmatrix}.$$

The third stage is to find another matrix N, which is the same size as T and is also upper triangular.* It can be derived from the known T matrix by applying the following equation:

$$n_{ii} = 1/t_{ii} \text{ and } n_{ij} = -\sum_{k=1}^{j-i} (n_{ik}\, t_{kj}/t_{jj}) \text{ where } j > i \tag{5.2}$$

In our example, we have therefore:

$$n_{11} = 1/t_{11} \qquad n_{22} = 1/t_{22} \qquad n_{33} = 1/t_{33}$$

$$n_{12} = -(n_{11}t_{12}/t_{22}) \qquad n_{13} = -(n_{11}t_{13} + n_{12}t_{23})/t_{33}$$

and $$n_{23} = -(n_{22}t_{23}/t_{33}).$$

---

* For those familiar with matrix algebra, N is in fact the inverse of T. The square root method may be used for finding the inverse.

This gives the following matrix for N:

$$N = \begin{bmatrix} 1.00 & -0.18 & -0.01 \\ 0.00 & 1.02 & -0.63 \\ 0.00 & 0.00 & 1.18 \end{bmatrix}.$$

The final stage is to produce the orthogonal factor matrix F by operating on the oblique factor matrix B using matrix N. The elements of F are obtained using the expression:

$$F_{ij} = \sum_{k=1}^{j} b_{ik}\, n_{kj} \tag{5.3}$$

For example:

$$\begin{aligned} F_{23} &= b_{21}n_{13} + b_{22}n_{23} + b_{23}n_{33} \\ &= 0.73(-0.01) + 0.04(-0.63) + 0.06(1.18) \\ &= 0.04 \end{aligned}$$

The complete orthogonal factor matrix is shown in Table 5.6.

TABLE 5.6 *Orthogonal factor matrix, showing percentage variance extracted by each factor (high loadings in italic)*

	Orthogonal factor			
Variable	F_1	F_2	F_3	Communalities h^2
1	*0.73*	0.09	−0.03	0.52
2	*0.73*	−0.09	0.04	0.54
3	0.06	*0.71*	0.12	0.52
4	0.16	*0.73*	−0.07	0.56
5	0.18	*0.69*	0.06	0.51
6	0.08	*0.58*	−0.13	0.36
7	0.12	*0.44*	*0.55*	0.51
8	0.03	*0.30*	*0.65*	0.51
% variance extracted	14%	27%	10%	51%

We can proceed from this orthogonal matrix to the factor pattern matrix P. However, before doing this it is useful to employ the orthogonal matrix to obtain the final residual matrix, since this will enable us to assess the adequacy of the factorisation procedure in this example. To the orthogonal matrix we apply equation (2.1) to obtain the expected values of the inter-correlations between variables and then subtract these values from the corresponding observed correlations to produce the residual values. Inspection of the size of the elements in the residual matrix will enable us to judge the worth of the factorisation. The resulting table of residuals is shown below:

	1	2	3	4	5	6	7	8
1	0.00	0.02	– 0.02	0.00	0.01	0.02	– 0.01	0.02
2		0.00	0.03	0.00	0.00	– 0.01	0.01	– 0.02
3			0.06	0.06	0.00	0.14	0.07	– 0.07
4				0.02	– 0.07	– 0.05	– 0.03	0.04
5					0.00	0.05	0.04	– 0.04
6						– 0.10	– 0.08	0.09
7							0.00	0.02
8								0.00

The values of the elements in this table are small enough to justify the conclusion that the factor analysis has been successful. Had the values been unacceptably large, it would have been an indication that our original hypothesis of a three-factor system was incorrect and the search for a fourth factor could be made by applying a factor analysis to the above matrix of residuals or by beginning again with a new four-factor hypothesis. We can now return to the derivation of the factor pattern matrix *P*.

5.5 Derivation of the factor pattern

We start with the orthogonal factor matrix *F*. The first step in obtaining *P* is to set up yet another matrix *L*,* which is obtained directly from matrix *N*; the columns of *N* become the rows of *L* and the rows of *N* the columns of *L*. We thus have:

$$L = \begin{bmatrix} 1.00 & 0.00 & 0.00 \\ -0.18 & 1.02 & 0.00 \\ -0.01 & -0.63 & 0.18 \end{bmatrix}.$$

We can use *L* to obtain the factor pattern matrix *P* (see p. 62) directly from the orthogonal factor matrix *F* by means of the following expression (which is analogous to (5.3)),

$$p_{ij} = \sum_{k=1}^{i} f_{ik} l_{kj}. \qquad (5.4)$$

* In matrix terminology *L* is the transpose of *N*.

This matrix is given in Table 5.5. As it happens, this factor pattern confirms our original hypotheses about which variables cluster together; if this was not the case, we could return to the orthogonal factor matrix F and rotate the factors or axes following the methods discussed in Chapter 3 to obtain simple structures. It follows that we could have started with a completely arbitrary hypothesis about the clustering of the variables and used the method as a preliminary to rotation.

The factor pattern is additionally useful in that it can be used to show the amount of variance extracted by each factor. It was mentioned in a previous chapter that the oblique factor loadings cannot be used to reproduce the original correlations between variables nor the communalities (i.e. the sums of squares of rows in the oblique factor structure does not equal the communalities). Moreover, the sums of squares of columns in the oblique factor structure does not tell us the amount of variance extracted by that factor. However, all of these things can be obtained once the factor pattern is known. Referring to the factor pattern, Table 5.5, let us consider variable 1. If this was an orthogonal factor matrix, the variance attributable to each factor would be given by the square of each factor loading respectively, e.g. the first factor p_1 contributes 0.71^2, i.e. 0.50 or 50 per cent of the variance of variable 1, whilst p_2 and p_3 contribute 0.11^2 and 0.04^2, or approximately 1 and 0.1 per cent, respectively. Moreover, the sum of these values, i.e. 51.1 per cent, should give the amount of common variance attributable to the three factors, i.e. should equal the communality. Where factors are oblique, as in the present factor pattern, some of the common variance of each variable is attributable to the joint influence of pairs of factors. For example, the joint contribution of p_1 and p_2 is given by $2(0.71 \times 0.11)\phi_{12}$, where ϕ_{12} is the correlation between p_1 and p_2, in this example equal to 0.18. Thus the joint contribution equals 0.03 or 3 per cent.

In general, the common variance V_i of variable i in an n factor solution is given by:

$$V_i = \sum_{j=1}^{n} p_{ij}^2 + 2 \sum_{\substack{r=1 \\ k=2}}^{k=n} \phi_{rk} p_{ir} p_{ik} \text{ where } r < k. \tag{5.5}$$

Where the first term on the right-hand side of equation (5.5) is the sum of the variances directly attributable to each factor, i.e. $0.71^2 + 0.11^2 + 0.04^2$, the second term is the total joint contribution of the factors.

For example, in this case for variable 1, the total variance is found by writing out (5.5) explicitly to give:

$$\begin{aligned} V_1 &= p_{11}^2 + p_{12}^2 + p_{13}^2 + 2\phi_{12}p_{11}p_{12} + 2\phi_{13}p_{11}p_{13} + 2\phi_{23}p_{12}p_{13} \\ &= 0.50 + 0.01 + 0.001 + 2 \times 0.18 \times 0.71 \times 0.11 \\ &\quad + 2 \times 0.11 \times 0.71 \times (-0.04) + 2 \times 0.53 \times 0.11 \times (-0.04) \\ &= 0.53 \end{aligned}$$

or 53 per cent.

This compares favourably (allowing for rounding errors) with the value for common variance of 0.54 or 54 per cent given in Table 5.6 for the orthogonal factor case. The communalities for each of the other variables can be obtained in precisely the same way and their total will be found to be approximately 51 per cent, the figure given for total common variance in Table 5.6. Thus, it can be observed that whereas 53 per cent of the common variance of variable 1 is attributable to the three common factors extracted by the analysis, 50 per cent of this is attributable directly to the first factor. This insight is gained by looking at the factor pattern and could not have been gained from the oblique factor structure. Thus, where an oblique solution is obtained to a problem, it is important that both the factor pattern *and* factor structure be presented.

It was indicated above that it was possible to derive the factor pattern directly from the oblique factor structure. We will now show how this can be done and how the residual matrix, which was calculated above using the orthogonal factor matrix, could have been obtained using the oblique factor structure and the factor pattern. Since this is possible, it is clear that all the results that can be obtained from the orthogonal factor matrix can be obtained from the other two matrices, and since the factor pattern can be obtained directly from the oblique factor structure (which is shown below), it is clear that the step of obtaining the orthogonal factor matrix from the oblique factor matrix is normally superfluous, and was shown merely for the sake of completeness. However, if rotation is deemed necessary when, for example, the factor pattern shows that our *a priori* hypothesis about the clustering of variables is unsatisfactory, then the rotation problem is greatly simplified by first obtaining the orthogonal factor matrix.

5.6 Obtaining the residuals from *B* and *P* directly

The expected values for the correlations between variables may be obtained from the formula:

$$r_{ij} = \sum_k b_{jk} p_{ik}; \qquad (5.6)$$

for example, the expected value of r_{45} is given by:

$$r_{45} = 0.18 \times 0.03 + 0.71 \times 0.78 + 0.43 \times (-0.08) = 0.53.$$

From these expected values the residuals can be obtained by subtraction in the usual way. The reader will notice that (5.6) is similar in appearance to the formula (2.1) used previously to calculate expected values in the case of orthogonal factor loadings. Indeed, it is a more general form of (2.1) since, when axes are orthogonal, the factor pattern is identical to the factor structure and both b_{ij} and p_{ij} are identical to F_{ij}. Hence (5.6) reduces to (2.1).

5.7 Obtaining *P* directly from *B*

The first step is to obtain a matrix, *M*, where elements are obtained from *N* by means of the formula:

$$m_{ij} = \sum_k n_{ik} n_{jk}. \tag{5.7}$$

Then the factor pattern matrix, *P*, is obtained from *M*, using the formula:

$$p_{ij} = \sum_k b_{ik} m_{kj}. \tag{5.8}$$

In our example, for instance:

$$\begin{aligned} m_{12} &= n_{11}n_{21} + n_{12}n_{22} + n_{13}n_{23} \\ &= 1.00 \times 0.00 + (-0.18 \times 1.02) + (-0.01) \times (-0.63) \\ &= -0.18. \end{aligned}$$

The complete matrix *M* can thus be calculated* as

$$M = \begin{bmatrix} 1.03 & -0.18 & -0.01 \\ -0.18 & 1.43 & -0.74 \\ -0.01 & -0.74 & 1.39 \end{bmatrix}$$

Thus, using equation (5.8), *P* can be calculated. For example:

$$\begin{aligned} p_{42} &= b_{41}m_{12} + b_{42}m_{22} + b_{43}m_{32} \\ &= 0.16 \times (-0.18) + 0.74 \times 1.43 + 0.34 \times (-0.74) \\ &= 0.78. \end{aligned}$$

The complete factor pattern matrix is given in Table 5.5.

5.8 Linkage analysis

At the beginning of this chapter it was stated that the multiple group method requires a determination of the number of factors likely to be present before the analysis can commence. This may be based on *a priori* knowledge or an informed guess. One systematic procedure which may be employed requires no *a priori* assumptions, and has the virtue of being simple. It is known as linkage analysis and is due to McQuitty [1957]. This procedure enables the researcher to establish

---

* In matrix terms *M* is the inverse of Φ, which is symmetrical, hence *M* is symmetrical. Therefore, the values below the principle diagonal need not be calculated.

automatically a system of clusters of variables on which the multiple group analysis may be then carried out.*

The linkage analysis procedure is as follows. Starting with a correlation matrix, such as that shown in Table 5.1, without the communalities, the following steps are carried out:

(i) underline the highest value in each column of the matrix;
(ii) select the highest value in the whole matrix; this gives the first two variables of the first cluster;
(iii) read along the rows corresponding to the two variables which emerged in step (ii), select any underlined values; the corresponding variables also belong to the first cluster;
(iv) read along the rows corresponding to the variables which emerged in step (iii) and again choose any underlined values; these variables again belong to the first cluster. Repeat this process until no further variables emerge. This completes the variables belonging to this cluster;
(v) excluding all the variables which fall into previous clusters, return to step (ii) and repeat the process.

If we apply the linkage analysis procedure above to the matrix of Table 5.1, the following three clusters emerge:

	variables
Cluster 1	3, 4, 5, 6
Cluster 2	1, 2
Cluster 3	7, 8

These are precisely the clusters which emerged from the multiple groups analysis and which we hypothesised.

The structure of the relationships between the variables within the clusters is usually shown diagrammatically, in this case, as in Figure 5.2, where the symbol $\rightleftarrows$ denotes a reciprocal relationship between two variables and are those emerging in step (ii), and the symbol $\rightarrow$ denotes that a variable at the

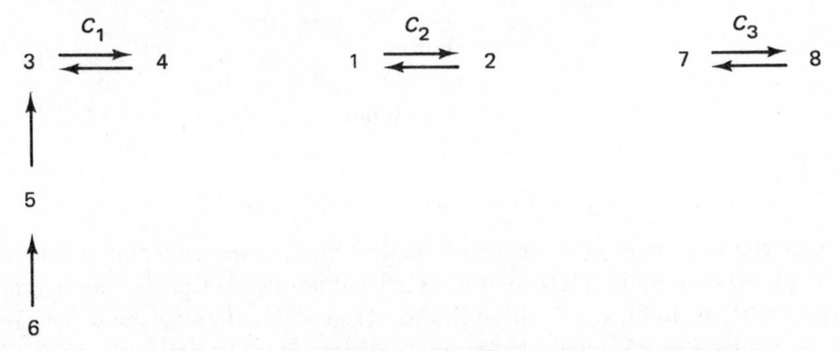

Fig. 5.2 *Diagrammatic illustration of linkage analysis*

*It should be noted that other more complex (and perhaps preferable) methods of cluster analysis exist.

tail of the arrow has its highest correlation with the variable at the head of the arrow (those emerging in steps (iii) and (iv)).

It may sometimes be possible to draw conclusions about the nature of the clusters from such a structural diagram. In factor analysis, when interpretation proves difficult, we might select the variable with the highest factor loading as being the best indicator of the nature of the factor. In linkage analysis we might select the variable which has the most arrows pointing to it. However, we present linkage analysis here mainly as a straightforward procedure for locating clusters prior to embarking on a multiple groups analysis.

It may be of interest to present an example of linkage analysis taken from the field of Social Studies.

This is an analysis of juvenile disturbance and uses Spearman rank correlations between 18 social variables carried out by Philip and McCulloch [1966]. Four clusters emerged and we present two of the more interesting below (the figures in brackets refer to the variables):

many children in care (5)
↓
high rate of self-poisoning (3) ⇆ many R.S.P.C.C. cases (18)
↑
many school absences (10)

Cluster 3

many stillbirths (12)
↓
low rate of adolescent psychiatric referrals (1)
↓
much juvenile deliquency (4) ⇄ much overcrowding (9)
↑
few owner-occupied homes (13)
↑
high infant mortality rate (11)

Cluster 4

Cluster 3 highlights the strong inter-connection between child cruelty and ultimate self-poisoning, a relationship which had not been suspected prior to the linkage analysis. In Cluster 4, although the relationship between delinquency and overcrowding is well known, what is not so obvious is the negative relationship which both of these have with adolescent psychiatric referrals, which is revealed by the linkage analysis. Thus, the analysis provides a starting point for further research by highlighting hitherto unsuspected relationships.

5.9 Flow-diagram

The steps taken in this chapter have been rather complex compared to previous ones, and it may be helpful to look at the flow-charts shown in Figure 5.3 in order to make clearer the relationships between the stages in the multiple groups analysis.

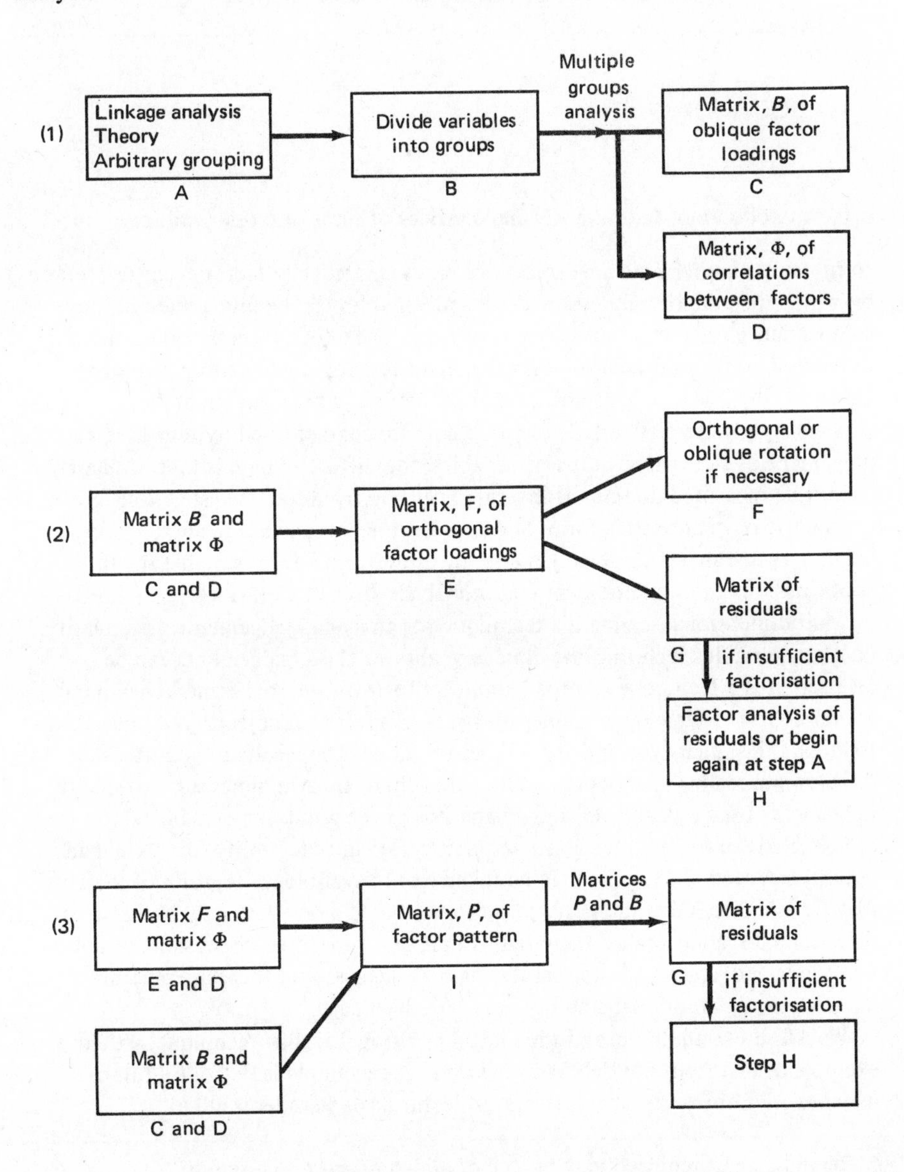

Fig. 5.3

6 Multidimensional Scaling

6.1 Introduction: factor-analysing matrices of sums of cross-products

In previous chapters, we have examined ways (specifically factor-analysis methods) by which the variation in a set of observed variables can be interpreted in terms of a dependence on a set of unobserved variables or factors. In most cases, obtaining the data in a form suitable to begin the analysis is reasonably straightforward. However, there is an important class of problem in which the initial data processing is more difficult. These problems are characterised by requiring a judgement of similarity* between variables; the difficulty here is that 'similarity' cannot be measured directly as the data resulting from such investigations is not in a metric or quantitative form. Multidimensional scaling is a technique which enables us to convert these non-metric measures into a form suitable for the application of those methods of factor analysis discussed previously.

Multidimensional scaling is a technique for analysing judgements of similarity between variables such that the dimensionality of those judgements can be assessed; it is a technique of broad applicability with the social and behavioural sciences as the following examples illustrate. A group of subjects could be asked to judge the similarity in quality of a set of goods. The results of a multidimensional scaling analysis may tell us that there are two dimensions of quality judgements used by subjects, and an analysis of the goods may enable us to attach labels to these dimensions, perhaps 'price' and 'durability' or 'style' and 'quality'. A second example may be taken from psychology. A group of subjects may be asked to judge similarity of 'personality' of a set of people. Multidimensional scaling analysis may tell us that there are three dimensions of judgement involved, and psychological testing may suggest that these are possibly 'extraversion', 'emotional stability' and 'intelligence'.

We will illustrate the use of the multidimensional scaling technique with an example drawn from the field of education. The example is fictitious, but one which would obviously be of interest to lecturers and students alike.

---

* The judgement required may be one of 'dis-similarity', 'closeness', 'relevance', etc. The essential point is that it is qualitative judgement that is required of the subject.

Example: 50 examiners were asked to assess the similarity in quality of seven essays by the method of triads (to be explained below). From these judgements, the 'distances' between the stimuli* were calculated. These were then converted into sums of cross-products, which were factor-analysed by the centroid method, which revealed that the essays could be represented in two dimensions as shown in Fig. 6.1. This suggests that the examiners assessed the essays on only two attributes and subsequent questioning and appropriate rotation of axes suggested that these might be tentatively labelled I, 'originality', and II, 'relevance'. The characteristics of each essay in respect of each of these dimensions can be seen from Fig. 6.1.

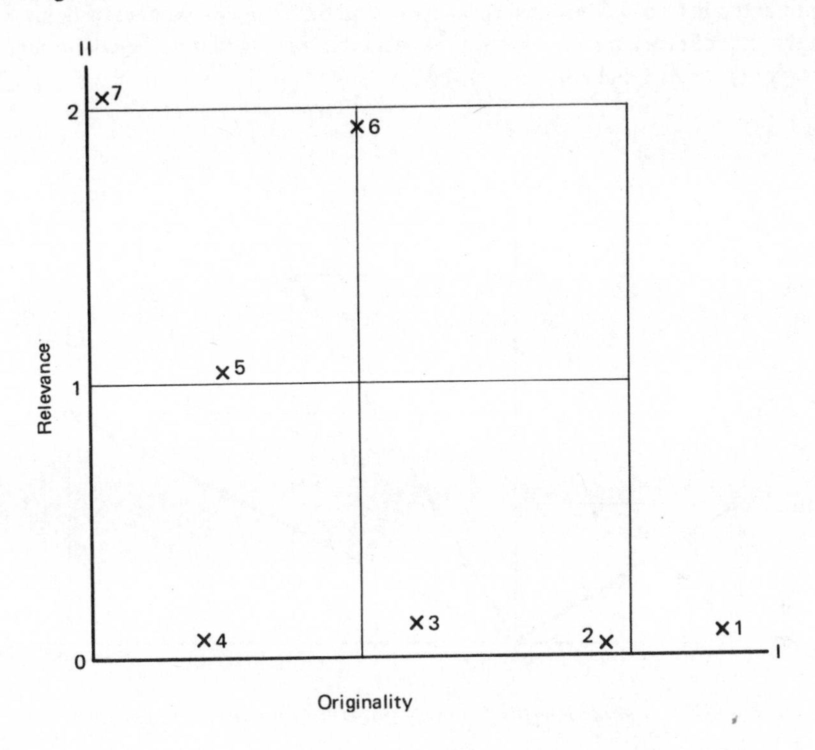

Fig. 6.1 *Two dimensions for the distances among seven essays*

The basic idea behind multidimensional scaling is quite straightforward. Suppose we have three variables or stimuli, P, Q and R. If we think of each as a point in space, then the problem becomes one of finding the *smallest* dimensionality of that space. Three points can always be represented in two-dimensional space, but sometimes in one. For example, if we know that the distances between the points are such that $PQ + QR = PR$, then the three points lie on a straight line and can

---

* The words 'stimuli' and 'variables' (in this example the 'essays'), are interchangeable.

therefore be represented on one dimension; if $PR < PQ + QR$ then the three points form a triangle and therefore two dimensions are required to represent them; if $PR > PQ + QR$, then the three points cannot exist in Euclidean space at all. Factor analysing sums of cross-products of distances enables us to say how many dimensions are required to represent the configuration of the points.

Before describing how the distances between stimuli are obtained, it is necessary to say a word or two about the factor analysis of sums of cross-products. In previous chapters we have shown how to factor analyse matrices of correlation coefficients, and matrices of sums of cross-products can be factored in precisely the same way.* For example, suppose the three points mentioned above can be represented in two-dimensional space, as in Fig. 6.2. The co-ordinate of point P on the first dimension is X_{1P} and on the second is X_{2P}; similarly, the co-ordinates of point Q are X_{1Q} and X_{2Q} respectively.

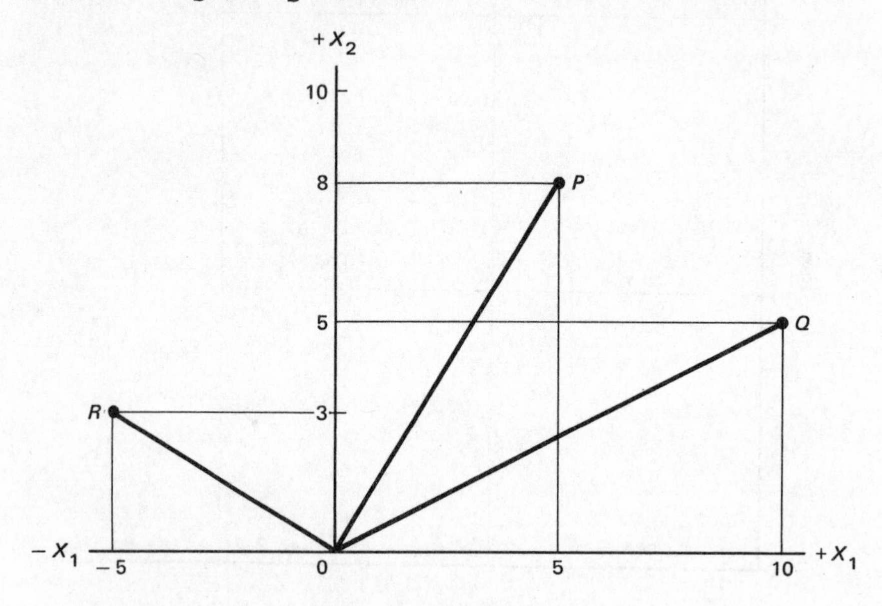

Fig. 6.2 *Vector lengths for three points in two-dimensional space*

The sum of cross-products for points P and Q, $\Sigma X_{iP} X_{iQ}$, $(i = 1, 2)$, is given by the product of the two X_1 co-ordinates plus the product of the two X_2 co-ordinates, i.e.

$$\Sigma X_{iP} X_{iQ} = (5 \times 10) + (8 \times 5) = 90.$$

* However, the variance of the variables affects the results of a factor analysis and therefore the scores on each variable should be normalised before analysing matrices of sums of cross-products. The calculation of correlation coefficients does this automatically, so that the problem does not arise when factor analysing matrices of correlations.

The full matrix of such sums of cross-products could be factor analysed by any of the methods described previously, and the original axes, X_1 and X_2, recovered.

However, in setting up a matrix of sums of cross-products prior to factor analysis in a multidimensional scaling problem, we are faced with the difficulty that the only information we have to begin with is a set of inter-point distances. However, examination of Fig. 6.2 will reveal that we can obtain the sums of cross-products, if we know the inter-point distances and the distances of each point from the origin. For example, in the case of points P and Q, if D_{PQ} is the distance between point P and point Q, then,

$$\begin{aligned} D_{PQ}^2 &= (X_{1Q} - X_{1P})^2 + (X_{2P} - X_{2Q})^2 \\ &= (X_{1Q}^2 + X_{2Q}^2) + (X_{1P}^2 + X_{2P}^2) - 2X_{1P}X_{1Q} - 2X_{2P}X_{2Q} \\ &= 0Q^2 + 0P^2 - 2\Sigma X_{iP}X_{iQ}, \end{aligned}$$

where $0Q$ and $0P$ are the distances of the points P and Q from the origin. Therefore:

$$\Sigma X_{iP}X_{iQ} = (0P^2 + 0Q^2 - D_{PQ}{}^2)/2 \qquad (6.1)$$

This equation is in fact quite general and applies with more than two dimensions. The only remaining problem is to fix an origin for the configuration of points. This is usually placed at the centroid of all the points since this would seem to be less prone to error than would other possibilities, such as choosing one of the points themselves as the origin. There are ways of calculating the distance of each point from centroid of the space, and we shall be making use of one such method later, but we shall leave the problem for now and refer the mathematically-inclined reader to the more detailed discussion in Torgerson [1958, Chapter II].

6.2 Estimating distances

The method of multidimensional scaling requires that distances be expressed in a ratio scale because the shape or configuration of the points only remains fixed and invariant when distances are measured on a ratio scale.* If distances are only measured on an interval scale, then it is possible that the points cannot be represented in Euclidean space at all. Hence, either we must obtain measures of distance on a ratio scale directly, as in the first method below, or obtain an interval-scale measure of the distances and then convert this to a ratio scale by the addition of a constant, as in the method used in the example below.

Direct ratio estimates

This method has the advantage over alternative methods described later of enabling distances to be assessed on a ratio scale directly. Subjects are asked to rate the similarity of all possible pairs of stimuli by giving a score of, say, 100, to two 'identical' stimuli and zero to pairs of stimuli which appear to be completely

---

*Readers unfamiliar with scales may refer to a suitable text, e.g. Guilford [1954].

dissimilar. This method has an apparent directness or simplicity but, in fact, it puts a considerable demand on subjects who may feel that they can assess similarity very roughly but who might find it difficult to make actual quantitative estimates. This makes the validity of the subject's judgements questionable and for this reason the method may be unsatisfactory in spite of being perhaps methodologically more direct, and we have not therefore employed it in our example.

Interval estimates

There are methods for producing similarity assessments directly on an interval scale, but they are too complex (in terms of the task besetting the subject) to be practically useful and they will not, therefore, be discussed further.

Ordinal estimates

This method has the disadvantage that it is less direct than the ratio method, producing in the first instance only information about the rank order of distances between pairs of stimuli. Thus a number of mathematical and psychological assumptions have to be made in order to convert this information into estimates of inter-stimulus distances on firstly an interval and subsequently a ratio scale. The advantage is that the task of judgement is very much simpler and more meaningful than it is with either of the two procedures above, and from a practical point of view it is far more useful.

One of the most frequent methods of producing ordinal estimates is the so-called method of triads. The subject is presented with three of the stimuli at a time and asked to say which two are the most similar and which two are the most dissimilar. If we label the three stimuli 1, 2 and 3, and 1 and 2 are judged most similar and 2 and 3 most dissimilar, then we put the three inter-stimulus distances into the rank order 12, 13 and 23. It will be shown in section 6.3 how this information can be used to generate proportions of the form ${}_kP_{ij}$, which is the proportion of times the distance between stimuli k and j, d_{kj}, is judged greater than the distance between stimuli k and i, d_{ki}. These proportions can be converted into a measure of the difference between two distances, ${}_kX_{ij} = d_{kj} - d_{ki}$, by a procedure based on Thurstone's [1927] law of comparative judgement. These ${}_kX_{ij}$'s can then be used to generate interval-scale estimates of inter-stimulus distances. How this is done will be described in a later section.

Conversion of proportions to interval-scale measures

There is not the space here to present a detailed discussion of Thurstone's model, and the interested reader is directed to Guilford [1954] for a clear exposition of the method.

Thurstone's model can be illustrated by considering the problem of a subject being asked to judge the heavier of two weights, W_1 and W_2, which are quite similar but where W_2 is in fact heavier than W_1. If the subject judges W_2 to be heavier 75 per cent of the time, this suggests that the intensity of his reaction or

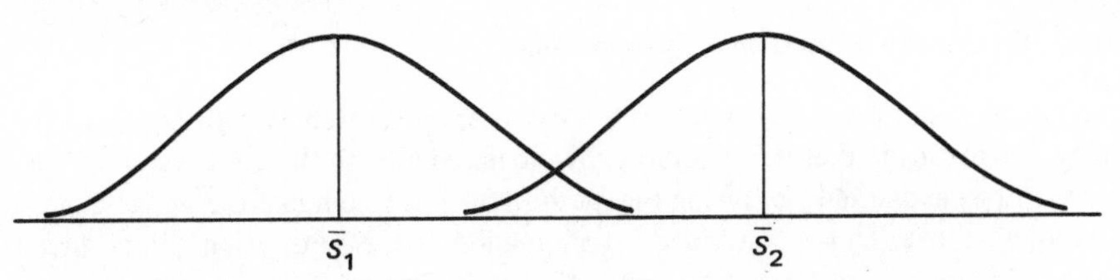

Fig. 6.3 *Distributions of reactions to two stimuli*

sensation to the weights varies over time but is on average greater in the case of W_2. Thurstone suggested that the reaction to a stimulus varies normally. Corresponding to W_1, therefore, is a measure of reaction intensity, S_1, which is distributed normally with a mean of $\bar{S}_1$. The distribution associated with W_2 has a mean of $\bar{S}_2$ and for this subject $\bar{S}_2 > \bar{S}_1$.

Thurstone made two simplifying assumptions: (i) the standard deviations, σ, of these distributions are equal, and (ii) S_1 and S_2 are independent. Since W_2 is not always judged the heavier, the distributions of S_1 and S_2 must overlap, as in Fig. 6.3. The distance between the two weights on a subjective scale of 'heaviness' is the distance between the means of the two distributions, $\bar{S}_2 - \bar{S}_1$.

It follows from the assumption above that:

$$\bar{S}_2 - \bar{S}_1 = z_{12} \times \sigma\sqrt{2},$$

where z_{12} is the normal deviate which cuts off 75 per cent of the unit normal distribution, i.e. $z_{12} = 0.67$.*

If we have n stimuli in a particular experiment, then we can obtain several ($\frac{1}{2}n\ (n + 1)$ to be exact) expressions for the distance in 'heaviness' between two stimuli, each of which will have the constant multiplier $\sigma \times \sqrt{2}$. Since we are only interested in measuring such distances on an interval scale and since an interval scale remains such when divided by a constant, this constant term can be dropped, giving, in general:

$$S_j - S_i = z_{ij}.$$

Instead of weights being assessed for 'heaviness', as in the above example, we are concerned at present with assessing differences in distances. Hence, we can substitute d_{kj} for S_j, d_{ki} for S_i, and ${}_kX_{ij}$, for z_{ij}, giving:

$${}_kX_{ij} = d_{kj} - d_{ki}.$$

The ${}_kX_{ij}$ mentioned previously in the section on ordinal estimates is thus seen to be the normal deviate that cuts off a proportion of the unit normal distribution equal to ${}_kP_{ij}$.

* If these notions are unfamiliar, any introductory statistics text should make them clear, e.g. Hoel [1971] and Yamane [1964].

6.3 Production of matrices of proportions

To return to our example on page 73, subjects are presented with three essays at a time (having made themselves thoroughly acquainted with the contents of all the essays first) and asked, following the method of triads, to make the judgements described above. This is continued for all possible triads (with seven stimuli this means 35 triads). Let us take the results from just one subject, which are set out in Table 6.1.

TABLE 6.1 *Judgements of similarity and dissimilarity between seven stimuli for one subject*

Triad	Most similar	Least similar	Rank order
123	(12)	(23)	d_{12}, d_{13}, d_{23}
124	(12)	(24)	d_{12}, d_{14}, d_{24}
125	(12)	(15)	d_{12}, d_{25}, d_{15}
.	.	.	.
.	.	.	.
.	.	.	.
567	(56)	(57)	d_{56}, d_{67}, d_{57}

Let ${}_kY_{ij}$ equal 1 if stimulus k is judged more similar to stimulus i than it is t stimulus j, and 0 otherwise. Then each triad (i, j, k) provides three values ${}_kY_{ij}$, ${}_jY_{ik}$ and ${}_iY_{jk}$. For example, for the first row in Table 6.1 we have ${}_1Y_{23} = 1$, ${}_3Y_{21} = 0$ and ${}_2Y_{13} = 1$. Next, we sum the values ${}_kY_{ij}$ across subjects to obtain values between 0 and 50, corresponding to the number of subjects, and divide by 50 to obtain values of ${}_kP_{ij}$, i.e. the proportion of times stimulus k is judged more simil: to stimulus i than it is to stimulus j, or the proportion of times d_{kj} is judged greater than d_{ki}.

6.4 Production of distances

From the normal probability tables, as described earlier, each proportion can be translated into its corresponding unit normal deviate, ${}_kX_{ij}$. Each ${}_kX_{ij}$ correspond to a difference between two distances and there will be 105 such differences (th from each triad). Each difference is an equation involving two unknown distanc and there are 21 unknowns altogether, so there is considerable over-identificati of the unknowns. We could choose any appropriate subset of the equations and solve for the distances, but our answers would depend on which subset we had chosen because the distances are determined from subject's judgements, which are prone to error. The solution to this problem, proposed by Torgerson [1958], is to choose those values for the distances which minimise the difference between the obtained numerical value of ${}_kX_{ij}$ and numerical values of $(d_{kj} - d_{ki})$ for all ${}_kX_{ij}$. In other words, the solution is to minimise the following function:

$$F = \sum_{k}^{n} \sum_{\substack{j \\ j \neq k}}^{n} \sum_{\substack{i \\ i \neq j \\ i \neq k}}^{n} ({}_kX_{ij} - (d_{kj} - d_{ki}))^2 .$$

Torgerson showed that this function is minimised when:

$$h_{jk} = \frac{1}{2} (m_{kj} + m_{jk} + 1_j + 1_k),$$

where

$h_{jk} = (d_{jk} - d_{..})$, the difference between the distance between two stimuli, d_{jk}, and the average distance between any two stimuli;

$$m_{jk} = \frac{1}{n-1} \sum_{\substack{i \\ i \neq k}}^{n} {}_kX_{ij}$$

$$= \frac{1}{n-1} \sum_{\substack{i \\ i \neq k}}^{n} (d_{ki} - d_{kj})$$ or the average distance between d_{ki} and d_{kj};

$$m_{jk} = \frac{1}{n-1} \sum_{\substack{i \\ i \neq j}}^{n} {}_jX_{ik} = \frac{1}{n-1} \sum_{\substack{i \\ i \neq j}}^{n} (d_{ji} - d_{jk})$$

or the average distance between d_{ji} and d_{jk};

$$1_j = \frac{1}{n} \sum_{i}^{n} m_{ij};$$

and $$1_k = \frac{1}{n} \sum_{i}^{n} m_{ik}.$$

We can show how the comparative distances, h_{jk}, are obtained from the values of ${}_kX_{ij}$ by constructing seven matrices $({}_1X_{ij}, {}_2X_{ij}, \ldots, {}_7X_{ij})$ as in Table 6.2.

TABLE 6.2 *Experimental differences in distances* $(d_{kj} - d_{ik})$

TABLE 6.2A $_1X_{ij}$

	1	2	3	4	5	6	7
1							
2			0.75	1.50	1.75	1.80	2.65
3		– 0.75		0.75	1.00	1.05	1.90
4		– 1.50	– 0.75		0.25	0.30	1.15
5		– 1.75	– 1.00	– 0.25		0.05	0.90
6		– 1.80	– 1.05	– 0.30	– 0.05		0.85
7		– 2.65	– 1.90	– 1.15	– 0.90	– 0.85	

TABLE 6.2B $_2X_{ij}$

	1	2	3	4	5	6	7
1			0.25	1.00	1.30	1.60	2.35
2							
3	– 0.25			0.75	1.05	1.35	2.10
4	– 1.00		– 0.75		0.30	0.60	1.35
5	– 1.30		– 1.05	– 0.30		0.30	1.05
6	– 1.60		– 1.35	– 0.60	– 0.30		0.75
7	– 2.35		– 2.10	– 1.35	– 1.05	– 0.75	

TABLE 6.2C $_3X_{ij}$

	1	2	3	4	5	6	7
1		– 0.50		– 0.50	– 0.05	– 0.50	1.05
2	0.50			0	0.45	1.00	1.55
3							
4	0.50	0			0.45	1.00	1.55
5	0.05	– 0.45		– 0.45		0.55	1.10
6	– 0.50	– 1.00		– 1.00	– 0.55		0.55
7	– 1.05	– 1.55		– 1.55	– 1.10	– 0.55	

TABLE 6.2D $_4X_{ij}$

	1	2	3	4	5	6	7
1		– 0.50	– 1.25		– 1.00	– 1.20	0.05
2	0.50		– 0.75		– 0.50	0.30	0.55
3	1.25	0.75			0.25	1.05	1.30
4							
5	1.00	0.50	– 0.25			0.80	1.05
6	0.20	– 0.30	– 1.05		– 0.80		0.25
7	– 0.05	– 0.55	– 1.30		– 1.05	– 0.25	

TABLE 6.2E $_5X_{ij}$

	1	2	3	4	5	6	7
1		−0.45	−1.05	−1.25		−1.35	−1.15
2	0.45		−0.60	−0.80		−0.90	−0.70
3	1.05	0.60		−0.20		−0.30	−1.10
4	1.25	0.80	0.20			−0.10	0.10
5							
6	1.35	0.90	0.30	0.10			0.20
7	1.15	0.70	0.10	−0.10		−0.20	

TABLE 6.2F $_6X_{ij}$

	1	2	3	4	5	6	7
1		−0.20	−0.55	−0.50	−1.40		−1.30
2	0.20		−0.35	−0.30	−1.20		−1.10
3	0.55	0.35		0.05	−0.85		−0.75
4	0.50	0.30	−0.05		−0.90		−0.80
5	1.40	1.20	0.85	0.90			0.10
6							
7	1.30	1.10	0.75	0.80	−0.10		

TABLE 6.2G $_7X_{ij}$

	1	2	3	4	5	6	7
1		−0.30	−0.85	−1.10	−2.05	−2.15	
2	0.30		−0.55	−0.80	−1.75	−1.85	
3	0.85	0.55		−0.25	−1.20	−1.30	
4	1.10	0.80	0.25		−0.95	−1.05	
5	2.05	1.75	1.20	0.95		−0.10	
6	2.15	1.85	1.30	1.05	0.10		
7							

Next we set up a matrix, M_{kj}, where each element is obtained from:

$$m_{kj} = \frac{1}{n-1} \sum_{\substack{i \\ i \neq k}}^{n} {}_kX_{ij}.$$

For example, $m_{12} = \frac{1}{6} ({}_1X_{22} + {}_1X_{32} + {}_1X_{42} + \ldots {}_1X_{72})$.

This expression is the sum of column 2 of the first matrix, ${}_1X_{ij}$, divided by six. In general, m_{kj} is the sum of column j of the kth matrix, divided by $n - 1$. M_{kj} is shown in Table 6.3.

TABLE 6.3 M_{kj}

	1	2	3	4	5	6	7
1		– 1.41	– 0.66	0.09	0.34	0.39	0.24
2	– 1.08		– 0.83	– 0.08	0.22	0.52	1.27
3	– 1.08	– 0.58		– 0.58	– 0.13	– 0.42	0.97
4	0.48	– 0.02	– 0.77		– 0.52	0.28	0.53
5	0.88	0.43	– 0.18	– 0.38		– 0.48	– 0.28
6	0.66	0.46	0.11	0.16	– 0.74		– 0.64
7	1.08	0.78	0.23	– 0.03	– 0.98	– 1.08	

Next, we obtain the matrix L_j, given in Table 6.4, from the averages of the columns of matrix M_{kj}. Thus, for example,

$$l_1 = \frac{1}{7} \sum_{k}^{n} m_{k1}.$$

TABLE 6.4 L_j

Averages of columns of matrix M_{kj}

1	2	3	4	5	6	7
0.28	– 0.05	– 0.30	– 0.12	– 0.26	0.01	0.44

Now, the comparative distances can be obtained from equation (6.2). For example,

$$h_{12} = \frac{1}{2}(m_{12} + m_{21} + l_1 + l_2)$$

$$\frac{1}{2}(-1.41 - 1.08 + 0.28 - 0.05)$$

$$= -1.13$$

These distances are set out in matrix H_{jk}, given in Table 6.5.

These distances are only comparative, not absolute measures, which is why some of them are negative. They must now be converted to true measures of distance (i.e. measures on a ratio scale with a true zero point) by the addition of an additive constant. The problem of finding the appropriate additive constant is one of finding the *smallest* number which preserves the Euclidean properties of the space containing the points. A space in which some of the distances between the points are negative is obviously non-Euclidean and therefore some constant must be added to the comparative distances. If the additive constant is made larger and larger, the dimensionality of the space becomes equal to the number of stimuli

TABLE 6.5 H_{jk}
Comparative distances

	1	2	3	4	5	6	7
1		– 1.13	– 0.38	0.37	0.62	0.67	1.52
2	– 1.13		– 0.88	– 0.13	0.17	0.47	1.22
3	– 0.38	– 0.88		– 0.88	– 0.43	0.12	0.67
4	0.37	– 0.13	– 0.88		– 0.63	0.17	0.42
5	0.62	0.17	– 0.43	– 0.63		– 0.73	– 0.53
6	0.67	0.47	0.12	0.17	– 0.73		– 0.63
7	1.52	– 0.22	0.67	0.42	– 0.53	– 0.63	

minus one. If the additive constant decreases then the dimensionality of the space may decrease, whilst at the same time the chances of the space becoming non-Euclidean increase. Hence the problem amounts to finding the smallest constant consistent with the preservation of Euclidean space, for this will give the smallest possible dimensionality for the space. Various methods of estimating the additive constant are given by Torgerson [1958], one of which will now be described.

Determining the additive constant

Consider any three stimuli, *i*, *j* and *k*. Suppose that the comparative distance, h_{ij}, is greater than the comparative distances h_{ik} and h_{jk}. Then the absolute distance, d_{ij}, is also greater than the absolute distances d_{ik} and d_{jk}. The absolute distances are obtained by adding a constant, A, to the corresponding comparative distances, i.e. $d_{ij} = h_{ij} + A$.

If we know that points 1, 2 and 3 lie on a straight line, then it is a simple matter to calculate the value of A. For example, let $h_{12} = 2$, $h_{23} = 1$ and $h_{13} = 4$. Since 1, 2 and 3 lie on a straight line, we know that,

$$d_{13} = d_{12} + d_{23}$$

Therefore $$(h_{13} + A) = (h_{12} + A) + (h_{23} + A).$$

Therefore $$h_{13} - h_{12} - h_{23} = A.$$

Therefore $$A = 4 - 2 - 1 = 1.$$

In general, let,

$$c_{ijk} = h_{ij} - h_{jk} - h_{jk}.$$

Each triad of points gives us the value of c_{ijk}. This will be largest for those three points which lie most nearly on a straight line. It seems reasonable to suppose that out of a total set of stimuli, at least three will fall approximately on a straight line. If this assumption is reasonable, then the largest value of c_{ijk} can be taken as the additive constant.

If we find more than one triad with large c_{ijk}'s then an average value can be taken. If four points lie roughly on a straight line, then we will have four high values of c_{ijk}, which can be averaged. Table 6.6 gives the values of c_{ijk} for the data in our example, calculated using equation (6.3).

TABLE 6.6 *Values of c_{ijk}*

ijk	c_{ijk}	ijk	c_{ijk}	ijk	c_{ijk}
123	1.63	156	− 0.03	267	1.13
124	1.63	157	1.23	345	0.53
125	1.53	167	1.33	346	0.23
126	1.03	234	1.63	347	0.63
127	1.23	235	1.33	356	0.93
134	1.63	236	0.83	357	1.63
135	1.23	237	1.23	367	0.73
136	0.23	245	0.63	456	1.43
137	0.83	246	− 0.77	457	1.53
145	0.13	247	0.23	467	0.13
146	− 1.17	256	0.43	567	0.03
147	− 0.17	257	1.53		

From Table 6.6 it appears that a one-dimensional subspace consisting of four points can be identified. The values are listed below:

1, 2, 3, 4

i, j, k	c_{ijk}
123	1.63
124	1.63
134	1.63
234	1.63

The fact that all four possible triads from points 1, 2, 3 and 4 give identical values of c_{ijk} is strong evidence that the four points lie on a straight line and therefore that c_{ijk} is indeed the appropriate additive constant; in addition, the fact that points 3, 5 and 7 also give the same value of c_{ijk} strengthens this assumption. In most cases, results would not be quite as good as in our example, i.e. the highest values of c_{ijk} will not be exactly equal, but a certain amount of error can be tolerated since the data is obtained from the fallible judgements of subjects. In such cases, the mean of the highest values of c_{ijk} will suffice.

Hence 1.63 is taken to be the additive constant and when this is added to the comparative distances, the absolute distances, d_{jk}, are obtained. These are set out in Table 6.7.

TABLE 6.7 D_{jk}
Absolute distances

	1	2	3	4	5	6	7
1		0.50	1.25	2.00	2.25	2.30	3.15
2	0.50		0.75	1.50	1.80	2.10	2.85
3	1.25	0.75		0.75	1.20	1.75	2.30
4	2.00	1.50	0.75		1.00	1.80	2.05
5	2.25	1.80	1.20	1.00		0.90	1.10
6	2.30	2.10	1.75	1.80	0.90		1.00
7	3.15	2.85	2.30	2.05	1.10	1.00	

6.5 Determination of sums of cross-products

As pointed out in section 6.1, the sums of cross-products of two stimuli P, Q can be obtained from equation (6.1):

$$\Sigma X_{iP}X_{iQ} = \frac{1}{2}(0P^2 + 0Q^2 - D_{PQ}{}^2),$$

or, if we let b_{jk} stand for the sum of cross-products for stimuli j and k, and d_{oj}, d_{ok} be the distances from the centroid to points j and k respectively, then we can re-write the equation as:

$$b_{jk} = \frac{1}{2}(d_{oj}^2 + d_{ok}^2 - d_{jk}^2). \tag{6.4}$$

Torgerson [1958, Chapter II] has shown that:

$$d_{oj}^2 + d_{ok}^2 = \frac{1}{n}\sum_{i}^{n} d_{jk}^2 + \frac{1}{n}\sum_{k}^{n} d_{jk}^2 - \frac{1}{n^2}\sum_{j}^{n}\sum_{k}^{n} d_{jk}^2 \cdot \tag{6.5}$$

This convenient formula means that we can work out the sums of cross-products directly from the inter-point distances without having to work out the distances from each point to the centroid. Combining equations (6.4) and (6.5) we obtain:

$$b_{jk} = \frac{1}{2}\frac{1}{n}(\sum_{j}^{n} d_{jk}^2 + \frac{1}{n}\sum_{k}^{n} d_{jk}^2 - \frac{1}{n_2}\sum_{j}^{n}\sum_{k}^{n} d_{jk}^2 - d_{jk}^2),$$

where

$$\frac{1}{n}\sum_{j}^{n} d_{jk}^2$$

is the average of the squared distances in column k of the D_{jk} matrix;

$$\frac{1}{n}\sum_{k}^{n} d_{jk}^2$$

is the average of the squared distances in column j of the D_{jk} matrix;

and
$$\frac{1}{n^2}\sum_j^n\sum_k^n d_{jk}^2$$

is the average of the square of each element in the D_{jk} matrix.

For example,

$$b_{12} = \frac{1}{2}\left(\frac{1}{7}\sum_j^7 d_{j2}^2 + \frac{1}{7}\sum_k^7 d_{1k}^2 - \frac{1}{49}\sum_j^n\sum_k^n d_{jk}^2 - d_{12}^2\right)$$

$$= \frac{1}{2}(2.69 + 3.73 - 2.71 - 0.25)$$

$$= 1.73.$$

The sums of cross-products are set out in Table 6.8.

TABLE 6.8 B_{jk}
Sums of cross-products

	1	2	3	4	5	6	7
1	2.38	1.73	0.62	− 0.40	− 1.11	− 0.87	− 2.33
2	1.73	1.34	0.60	− 0.05	− 0.72	− 0.95	− 1.95
3	0.62	0.60	0.43	0.35	− 0.28	− 0.73	− 0.99
4	− 0.40	− 0.05	0.35	0.83	0.15	− 0.62	− 0.24
5	− 1.11	− 0.72	− 0.28	0.15	0.47	0.42	1.08
6	− 0.87	− 0.95	− 0.73	− 0.62	0.42	1.19	1.54
7	− 2.33	− 1.95	− 0.99	− 0.24	1.08	1.54	2.90

6.6 Factor analysis of sums of cross-products

The B_{jk} matrix may be factor analysed by any method, but we have carried it out by the centroid method. The workings of the analysis proceed exactly as described in Chapter 3, and they are set out below without extensive comment.

Determination of which variables to reflect

We use the systematic procedure described in Chapter 3 to determine which variables to reflect. Ignoring the diagonal entries, let A_i be the sum of the positive elements in column i, and B_i be the sum of the negatives. Then recall that $(B_i - A_i)$ is a measure of the extent to which the sum of the elements is increased by the reflecting variable i. (See Table 6.9.)

TABLE 6.9 *Values of $(B_i - A_i)$ for several reflections*

	stimulus						
	1	2	3	4	5	6	7
A_i	2.35	2.33	1.57	0.50	1.65	1.96	2.62
B_i	4.71	3.67	2.00	1.31	2.11	3.17	5.51
$(B_i - A_i)$	2.36	1.34	0.43	0.81	0.46	1.21	2.89
				Reflect 7			
A_i	4.68	4.28	2.56	0.74	0.57	0.42	5.51
B_i	2.38	1.72	1.01	1.07	3.19	2.75	2.62
$(B_i - A_i)$	–	–	–	0.33	2.62	2.33	–
				Reflect 5			
A_i	5.79	5.00	2.84	0.59	3.19	0	6.59
B_i	1.27	1.00	0.73	1.22	0.57	3.17	1.54
$(B_i - A_i)$	–	–	–	0.63	–	3.17	–
				Reflect 6			
A_i	6.66	5.95	3.57	1.21	3.61	3.17	8.13
B_i	0.40	0.05	0	0.60	0.15	0	0
$(B_i - A_i)$	–	–	–	–	–	–	–
				STOP			

Production of first residual matrix

The steps up to the production of the first residual matrix are illustrated in Tables 6.10, 6.11 and 6.12.

TABLE 6.10 B_{jk} *(5, 6, 7)*
The B_{jk} matrix after reflection

	1	2	3	4	5	6	7	Row total	F_1 loading
1	2.38	1.73	0.62	– 0.40	1.11	0.87	2.33	8.64	1.32
2	1.73	1.34	0.60	1.05	0.72	0.95	1.95	7.24	1.11
3	0.62	0.60	0.43	0.35	0.28	0.73	0.99	4.00	0.62
4	– 0.40	– 0.05	0.35	0.83	– 0.15	0.62	0.24	1.44	0.22
5	1.11	0.72	0.28	– 1.15	0.47	0.42	1.08	(–) 3.93	– 0.60
6	0.87	0.95	0.73	0.63	0.42	1.19	1.54	(–) 6.32	– 0.97
7	2.33	1.95	0.99	0.24	1.08	1.54	2.90	(–)11.03	– 1.69
								42.60	

TABLE 6.11 E_1
Expected values

	1	2	3	4	5	6	7
1	1.75	1.47	0.81	0.29	– 0.80	– 1.28	– 2.24
2	1.47	1.23	0.70	0.24	– 0.67	– 1.07	– 1.87
3	0.81	0.70	0.38	0.14	– 0.37	– 0.59	– 1.03
4	0.29	0.24	0.14	0.05	– 0.13	– 0.21	– 0.37
5	– 0.80	– 0.67	– 0.37	– 0.13	0.36	0.58	1.02
6	– 1.28	– 1.07	– 0.59	– 0.21	0.58	0.94	1.64
7	– 2.24	– 1.87	– 1.03	– 0.37	1.02	1.64	2.86

TABLE 6.12 R_1
Residual values

	1	2	3	4	5	6	7
1	0.63	0.26	– 0.19	– 0.69	– 0.31	0.41	– 0.09
2	0.26	0.11	– 0.10	– 0.29	– 0.05	0.12	– 0.08
3	– 0.19	– 0.10	0.05	0.21	0.09	– 0.14	0.04
4	– 0.69	– 0.29	0.21	0.78	0.28	– 0.41	0.13
5	– 0.31	– 0.05	0.09	0.28	0.11	– 0.16	0.06
6	0.41	0.12	– 0.14	– 0.41	– 0.16	0.25	– 0.10
7	– 0.09	– 0.08	0.04	0.13	0.06	– 0.10	0.04

Determination of which variables to reflect

The requisite calculations for this are laid out in Table 6.13.

TABLE 6.13 *Values of $(B_i - A_i)$ for several reflections*

	1	2	3	4	5	6	7
A_i	0.67	0.38	0.34	0.62	0.43	0.53	0.23
B_i	1.28	0.52	0.43	1.39	0.52	0.81	0.27
$(B_i - A_i)$	0.61	0.14	0.09	0.77	0.09	0.28	0.04
				Reflect 4			
A_i	1.36	0.67	0.13	1.39	0.15	0.94	0.10
B_i	0.59	0.23	0.64	0.62	0.80	0.40	0.40
$(B_i - A_i)$	–	–	0.51	–	0.65	–	0.30
				Reflect 5			
A_i	1.67	0.72	0.04	1.67	0.80	1.10	0.04
B_i	0.28	0.18	0.73	0.34	0.15	0.24	0.46
$(B_i - A_i)$	–	–	0.69	–	–	–	0.42

TABLE 6.13 *Values of $(B_i - A_i)$ for several reflections continued*

	1	2	3	4	5	6	7
				Reflect 3			
A_i	1.86	0.82	0.73	1.88	0.89	1.24	0
B_i	0.09	0.08	0.04	0.18	0.06	0.10	0.50
$(B_i - A_i)$	–	–	–	–	–	–	0.50
				Reflect 7			
A_i	1.95	0.90	0.77	2.01	0.95	1.34	0.50
B_i	0	0	0	0	0	0	0
$(B_i - A_i)$	–	–	–	–	–	–	–
				STOP			

Production of second residual matrix

This is illustrated in Tables, 6.14, 6.15 and 6.16.

TABLE 6.14 R_1 (3, 4, 5, 7)
First residual matrix after reflection

	1	2	3	4	5	6	7	row total	F_2 loading
1	0.63	0.26	0.19	0.69	0.31	0.41	0.09	2.58	0.79
2	0.26	0.11	0.10	0.29	0.05	0.12	0.08	1.01	0.33
3	0.19	0.10	0.05	0.21	0.09	0.14	0.04	(–) 0.82	– 0.22
4	0.69	0.29	0.21	0.78	0.28	0.41	0.13	(–) 2.79	– 0.88
5	0.31	0.05	0.09	0.28	0.11	0.16	0.06	(–) 1.06	– 0.33
6	0.41	0.12	0.14	0.41	0.16	0.25	0.10	1.59	0.50
7	0.09	0.08	0.04	0.13	0.06	0.10	0.04	(–) 0.54	– 0.20
								10.39	

TABLE 6.15 E_2
Expected values

	1	2	3	4	5	6	7
1	0.64	0.25	– 0.20	– 0.69	– 0.26	0.39	– 0.13
2	0.25	0.10	– 0.08	– 0.27	– 0.10	0.15	– 0.05
3	– 0.20	– 0.08	0.06	0.22	0.08	– 0.13	0.04
4	– 0.69	– 0.27	0.22	0.75	0.28	– 0.43	0.15
5	– 0.26	– 0.10	0.08	0.28	0.11	– 0.16	0.06
6	0.39	0.15	– 0.13	– 0.43	– 0.16	0.24	– 0.08
7	– 0.13	– 0.05	0.04	0.15	0.06	– 0.08	0.03

TABLE 6.16 R_2
Second residuals

	1	2	3	4	5	6	7
1	−0.01	0.01	0.01	0.00	−0.05	0.02	0.04
2	0.01	0.01	−0.02	−0.02	0.05	−0.03	−0.03
3	0.01	−0.02	−0.01	−0.01	0.01	−0.01	0.00
4	0.00	−0.02	−0.01	0.03	0.00	0.02	−0.02
5	−0.05	0.05	0.01	0.00	0.00	0.00	0.00
6	0.02	−0.03	0.01	0.02	0.00	0.01	−0.02
7	0.04	−0.03	0.00	−0.02	0.00	−0.02	0.01

Since the elements of matrix R_2 are so small, we conclude that the seven points can be adequately represented in two-dimensional space. The factor loadings are set out in Table 6.17. These were then subjected to a transformation that (*a*) rotated the axes, and (*b*) moved the origin; this had the effect of (*a*) making all the factor loadings positive, and (*b*) gave as many low-factor loadings as possible. This was done, in accordance with the discussion of rotation of axes given in Chapter 3, so as to make the results more amenable to psychological interpretation. The transformed factor loadings are given in Table 6.17 and the configuration of points presented diagrammatically in Fig. 6.1. The transformation employed on the factor loadings was firstly to rotate the axes through 56.3°,

$$F_1' = F_1 \cos 56.3^\circ + F_2 \sin 56.3^\circ$$

$$F_2' = F_2 \cos 56.3^\circ - F_1 \sin 56.3^\circ$$

and secondly to shift the origin to remove all negative factor loadings,

$$F_1'' = F_1' + 1.10$$

$$F_2'' = F_2' + 0.74$$

TABLE 6.17 *Original and transformed factor loadings for seven essays on two dimensions*

	1	2	3	4	5	6	7
F_1	1.32	1.11	0.62	0.22	−0.60	−0.97	−1.69
F_2	0.79	0.33	−0.22	−0.88	−0.33	0.50	−0.20
F_1'	1.38	0.88	0.16	−0.61	−0.60	−0.12	−1.10
F_2'	−0.66	−0.74	−0.64	−0.67	0.32	1.08	1.29
F_1''	2.48	1.98	1.26	0.49	0.50	0.98	0
F_2''	0.08	0	0.10	0.05	1.06	1.82	2.03

6.7 Interpretation of results

So far, the analysis has revealed that only two dimensions are required to explain the configuration of points and has given the factor loadings or co-ordinates of each point in respect of those two dimensions. We see from Table 6.17 that points 1, 2 and 3 have high loadings on Factor I and very low loadings on Factor II. Point 4 has a moderate loading on Factor I and a very low loading on Factor II. Point 7 has a high loading on Factor II and a zero loading on the first factor. Points 5 and 6 occupy somewhat intermediate positions.

It appears that the 50 examiners have only made use of two attributes in judging the quality of the seven essays. However, we must now give a meaning to each of these attributes or dimensions; methods for doing this have been described in Chapter 3. In this case, the most obvious approach would be to discuss with each examiner what it is that essays, which have been discovered to be most similar in the analysis, appear to have in common. A more systematic approach might be to present each examiner with three essays at a time, two of which are known from the analysis to be more similar to each other than either is to the third. In our example, an examiner might be presented with essays 1, 2 and 6 and asked to describe in a word or short phrase what it is that makes 1 and 2 more similar to each other than either is to essay 6. Typical answers might be 'more relevant', 'clearly argued', 'original', 'length' and so on. From many such responses one might expect two such characteristics to emerge as predominant. In our fictitious example, we have named the two dimensions 'relevance' and 'originality'.

We can now go on to consider the extent to which our attributes appear to correlate, as described in Chapter 4. In our fictitious example, 'originality' appears to be relatively uncorrelated with 'relevance'. Thus the analysis gives us an instant insight into not only the dimensionality of the judgements made by examiners but also reveals something about the stimuli themselves, i.e. that people who use an original approach in writing essays are not the same as those whose essays have a high degree of relevance. This, of course, is a fictitious example and we mention it only to show the kind of insights that might possibly be obtained from the results of a multidimensional scaling analysis, i.e. insights into the nature of judgements and into the nature of stimuli employed.

6.8 Assumptions and difficulties of the method

One of the assumptions that has to be made when distances are initially estimated on an ordinal scale, as in our worked example, is the validity of Thurstone's law of comparative judgement. It is difficult to know for certain whether the law is valid in any situation, but it has certainly held up well in situations involving stimuli that vary along known, physical dimensions. For example, Torgerson describes an experiment in which subjects are asked to judge the similarity of coloured chips which varied in brightness and saturation. Analysis revealed two dimensions corresponding well to the known dimensions, thus giving the method considerable plausibility.

However, where the dimensions involved are not physical ones, no such direct check is possible. Here, the only check available to the investigator is to repeat the experiment with some additional stimuli. If these stimuli involve no new dimensions, then the inter-stimulus distances of the stimuli which have been used in the previous study should remain essentially unchanged.

Another problem concerns the difficulty of the task confronting the subjects. Some subjects, and even some investigators, find the instructions to assess the similarity of stimuli a little confusing, especially where variation along several dimensions exists. The problem can be minimised by ensuring that the instructions do not require too much of the subjects, which is the case when direct ratio-scale estimates of distance are required. But, ultimately, the method is justified by the results it produces. The meaningfulness of the dimensions arising from the method can only be gauged from the results of further research, i.e. they should be usable as independent variables in subsequent experiments.

For example, we could postulate that essay marks are dependent on the attributes we have discovered from the previous analysis. Then essays which score more highly on attribute A (with attribute B held constant) should also get a higher mark; the same can be said with the attributes reversed. A more ambitious hypothesis might involve the devising of an equation expressing the essay mark as a function of the factor loadings on the two attributes. The constants in such an equation might be estimated using some of the essays, and then the equation may be used to predict the marks of the remaining essays. If such experimental hypotheses are found to be supported, then confidence in the meaningfulness of the dimensions derived from the analysis can be increased.

The maximum number of dimensions that can be obtained is $n-1$ where n is the number of stimuli. If subjects find the judgement task too difficult then the data derived from their judgements will be prone to a great deal of error, increasing the dimensionality of the space required to represent the points. Hence, if the analysis enables us to conclude that fewer than $n-1$ dimensions are necessary, then we are entitled to believe that subjects do find the task a meaningful one. Our confidence in this belief will be greater the fewer the number of dimensions required.

The number of stimuli employed in our worked example was seven. It is advisable to have as many stimuli as conditions will permit, because this decreases the chances of failing to detect significant dimensions, which may be missed with a smaller number. As the number of stimuli increases, so the number of triads presented to the subjects increases; with large numbers of stimuli, the method of triads becomes impracticable. Other more convenient methods are available, and the interested reader is referred to the discussion in Coombs [1964].

In the method of obtaining interval scale estimates of distance from ${}_kP_{ij}$'s, no such proportions must be zero or unity, since no corresponding values of ${}_kX_{ij}$ exist. The main safeguard against this is to use a sufficiently large number of subjects, but if one stimulus should be too extreme, then this may not be practicable. The only answer then is to delete that extreme stimulus from the total set of stimuli. A preliminary pilot study should enable the investigator to avoid this problem.

Multidimensional scaling analysis is normally too complex a task to undertake without the aid of a computer and, since this is the case, the principal component method of factorising may be a more appropriate technique than the centroid method. The principal component method has a built-in check of the Euclidean properties of the space containing the points, i.e. all the latent roots must be positive or zero; negative latent roots imply that the space is non-Euclidean. Hence the additive constant can be obtained by iteration and checking on the latent roots, although iterative procedures are too lengthy to employ without the aid of a computer.

6.9 Scaling with unknown distance functions

Some more recent methods of multidimensional scaling have side-stepped the problem of estimating distances prior to analysis (e.g. Kruskal [1964]). These methods start with information about the rank ordering of stimuli; this is taken to be an approximation of the distances, and the approximation is then altered according to certain mathematical criteria. The methods and rationale behind them are very complex and well beyond the scope of this text, so no further discussion of them can be given here. The whole field of multidimensional scaling is a rapidly developing one, and many more methods of obtaining data from subjects, data processing and the analysis of this data exist. All we have been able to attempt in this brief treatment is to introduce the reader to some of the underlying notions and potential applications of the method.

6.10 An example from the literature

We conclude this chapter with an example of multidimensional scaling taken from the field of social planning. In particular, a study by Dobson [1974] of public attitudes towards a new transport system, the purpose of which was to discover what underlying dimensions existed in consumer preferences towards various attributes of the proposed system. It was hoped that the study would provide information potentially useful in the planning and implementation of the system under review or of similar system - environmental combinations.

Subjects were asked to make comparisons between all pairs of 32 design attributes of a demand-responsive public transport system in a low-density suburb of a large metropolitan area. These attributes included such items as:

(1) A shorter time spent travelling in the vehicle.
(2) A shorter time spent waiting to be picked up.

(3) Arriving at destination punctually.
(4) Ability to adjust the amount of light, heat, air and sound around you in the vehicle.

and so on, down to,

(32) Convenient method of paying your fare.

Since comparisons between all possible pairs of attributes would have yielded 496 choices, which was thought too large a number for respondents to handle, the list was split into smaller sub-groups each relating to some specific area, e.g. vehicle design. Respondents were then asked to choose between all pairs in each sub-group and the replies from each aggregated into a complete choice vector over all sub-groups.

Principal components analysis was then applied to the aggregated choice vectors and six components were extracted and rotated using the Varimax procedure. The first component is interpreted as being related to 'level of service' attributes at one extreme and to a conglomerate of attributes (aesthetic, amenity, social interaction) at the other. The remaining components are difficult to interpret and the reader is referred to the original paper for the author's comments in this respect.

7 Discriminant Analysis

7.1 Introduction

Previous chapters have dealt with techniques (mainly factor analysis) which are useful in the analysis of the relationships between a set of variables. The variables were in fact the results of several measures taken on a single group of objects or persons. In the following chapter we will examine a rather different problem, that in which several groups of objects or persons are measured on the same set of variables. The principal question to be answered is can the groups be differentiated on the basis of the measures obtained from the set of variables? This chapter describes a number of ways of approaching this and other related problems.

7.2 Discriminant function analysis: the case of two groups

Suppose we have two groups of politically aware people, for example 60 Democrats and 80 Republicans, and that each individual is measured on three variables: X_1, own income; X_2, years of education; and X_3, father's income. We want to know whether the measurements we obtain on the three variables can be used as a means of discriminating between Democrat and Republican supporters and, if so, what is the most efficient way of doing this? Discriminant function analysis reduces these three variables to a single one which is compounded of the original three. Y_i, the ith individual's score on this discriminant function, is obtained by multiplying his or her score on each of the three variables by an appropriate weight, w_i, as follows:

$$Y_i = \sum_i w_i X_i = w_1 X_1 + w_2 X_2 + w_3 X_3. \qquad (7.1)$$

The weights are chosen in such a way that the discriminant function scores, the Y's, for each group are maximally separated from one another, i.e. so that $\bar{Y}_1$ and $\bar{Y}_2$, the average discriminant function scores for groups one and two respectively, are as far apart as possible.

The method requires that we can assume that the variables within each group have a multinormal distribution. In addition, we have to assume that the variances of a particular variable for each group are similar and that the covariances between pairs of variables are likewise similar for each group. If these assumptions are satisfied, then the discriminant function scores for each group will be normally distributed and have equal variances, a fact which makes possible a number of interesting interpretations once the weights have been obtained.

7.3 Computing the weights, w_i

Table 7.1 gives the means on each variable for the two groups in our example; n_i is the number of people in each sample group, and d_i is the difference between means for variable *i*.

TABLE 7.1 *Means of scores on the three variables for Republicans and Democrats*

Variable	Groups		
	1 Republican	2 Democrat	d_i
Own income, X_i	34	30	4
Years of education, X_2	12	10	2
Father's income, X_3	35	25	10
Number in each group, n_i	80	60	

Let us consider group one, Republicans, first. The within-groups sums of squares for each variable are obtained; for example, for X_1 (own income), the value is:

$$\sum (X_1 - \bar{X}_1)^2 = \sum x_1^2.$$

It is convenient to add a second subscript to denote that this value is obtained for the first group; the above term thus becomes Σx_{11}^2. Next, the sums of cross-products for each variable are obtained. Thus the sum of cross-products for variables 1 and 2 is:

$$\sum (X_1 - \bar{X}_1)(X_2 - \bar{X}_2) = \sum x_{11} x_{21}$$

The values obtained can be conveniently represented in the form of a matrix, S_1 (the subscript indicating group one), where the diagonal entries represent the sums of squares of variables and the off-diagonal entries the sums of cross-products:

$$S_1 = \begin{bmatrix} \sum x_{11}^2 & \sum x_{11}x_{21} & \sum x_{11}x_{31} \\ & \sum x_{21}^2 & \sum x_{21}x_{31} \\ & & \sum x_{31}^2 \end{bmatrix}.$$

Turning now to group two (Democrat), a similar matrix S_2 can be constructed. Providing our assumption that the variances and covariances for each group are similar, we can add together corresponding elements of S_1 and S_2 to obtain a pooled within-groups sum of squares and sums of cross-products matrix, W. These operations, with actual data inserted from our example, are shown below:

$$\overset{S_1}{\begin{bmatrix} 1103.172 & 315.192 & 393.990 \\ & 315.192 & 236.394 \\ & & 1575.960 \end{bmatrix}} + \overset{S_2}{\begin{bmatrix} 828.828 & 236.808 & 296.010 \\ & 236.808 & 177.606 \\ & & 1184.040 \end{bmatrix}}$$

$$= \overset{W}{\begin{bmatrix} 1932 & 552 & 690 \\ & 552 & 414 \\ & & 2760 \end{bmatrix}}.$$

The matrix, W, is now divided by the within-groups degrees of freedom (in general

$$\sum_i^k n_i - k,$$

where n_i is the number of persons or objects in the ith group and k is the number of groups), to give the within-groups covariance matrix, V. In our example, the number of degress of freedom is 138. The diagonal entries in V are the within-groups variance of each variable and the off-diagonal entries the within-groups covariance between pairs of variables. The covariance matrix for our example is:

$$V = \begin{bmatrix} 14 & 4 & 5 \\ & 4 & 3 \\ & & 20 \end{bmatrix}.$$

The inverse, U, of this matrix is now calculated using the square-root method (see Chapter 1). We find that,

$$U = \begin{bmatrix} 0.102 & -0.094 & -0.012 \\ -0.094 & 0.367 & -0.032 \\ -0.012 & -0.032 & 0.058 \end{bmatrix}.$$

The weights w_i can now be obtained from the formula:

$$w_i = \sum_{j=1}^{m} u_{ij} d_j,$$

where m is the number of variables. For example,

$$\begin{aligned} w_1 &= u_{11}d_1 + u_{12}d_2 + u_{13}d_3 \\ &= (0.102 \times 4) + (-0.094 \times 2) + (-0.012 \times 10) \\ &= 0.100 \end{aligned}$$

Similarly, $w_2 = 0.038$ and $w_3 = 0.468$.

7.4 Obtaining discriminant function scores

The average discriminant function scores for group one are obtained from equation (7.1) by multiplying the means of each variable by their corresponding weights and summing these products. Thus,

$$\overline{Y}_1 = \sum_{i=1}^{m} w_i \overline{X}_{i1}, \qquad (7.2)$$

which in our example gives:

$$\overline{Y}_1 = (0.100 \times 34) + (0.038 \times 12) + (0.468 \times 35) = 20.236.$$

Similarly, the average for the second group is given by:

$$\overline{Y}_2 = \sum_{i=1}^{m} w_i \overline{X}_{i2} = 15.08.$$

Equation (7.1) can be used to calculate a person's score on Y from his score on the three variables. Thus, a person who has the following scores, $X_1 = 20$, $X_2 = 8$ and $X_3 = 30$, would have a score on the discriminant function Y, given by,

$$Y = (0.100 \times 20) + (0.038 \times 8) + (0.468 \times 30) = 16.344.$$

How we decide into which group this person should be assigned (Democrat or Republican) on the basis of this score will be considered later.

7.5 Testing the significance of the discriminant function

Having obtained a discriminant function we are obviously interested to know if it can *significantly* distinguish between the two groups. The first step is to calculate the within-group variance of the discriminant function scores, V_y, given by:

$$V_y = \sum_{i=1}^{m} w_i d_i. \tag{7.3}$$

In our example, $V_y = (0.100 \times 4) + (0.038 \times 2) + (0.468 \times 10) = 5.156$

It is convenient for the purposes of later calculations to re-scale the discriminant function scores so that the within-group variance equals unity. This is achieved by dividing the weights by the square root of V_y to obtain a new set of weights. For example,

$$w_1 = \frac{0.100}{\sqrt{5.156}} = 0.044, \qquad w_2 = 0.017, \qquad w_3 = 0.206$$

Using the new scale, the mean value of Y for each group is given by (7.2), i.e.

$$\bar{Y}_1 = 8.910 \text{ and } \bar{Y}_2 = 6.640$$

Fig. 7.1 gives a diagrammatic representation of what the discriminant analysis has produced.

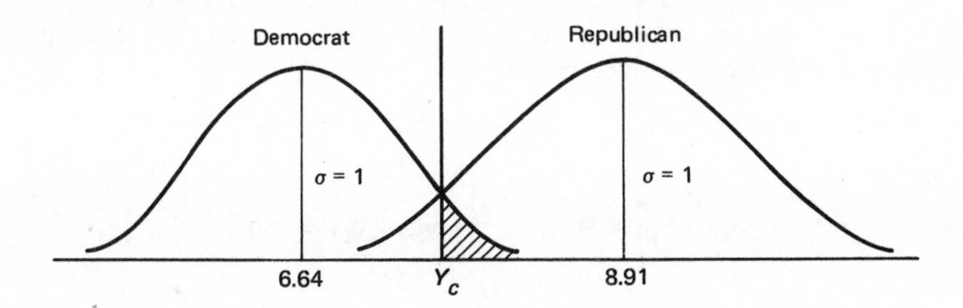

Fig. 7.1 *Distributions of discriminant function scores for the two groups*

How well are the two groups differentiated by the discriminant function? A quick, approximate method of deciding whether the two groups can be significantly discriminated by the three variables is to use the well-known test of significance for the difference between the means of two normal distributions, i.e. obtain the score

$$z = (\bar{Y}_1 - \bar{Y}_2)/\sqrt{(\delta(Y_1)/n_1) + (\delta(Y_2)/n_2)},$$

and look up the value of z in the normal probability tables. In our example, owing to the re-scaling of Y, $\delta(Y_1) = \delta(Y_2) = 1$. Hence

$$z = (\bar{Y}_1 - \bar{Y}_2)/\sqrt{\frac{1}{n_1} + \frac{1}{n_2}} = (8.91 - 6.64)/\sqrt{\frac{1}{80} + \frac{1}{60}} = 13.275.$$

Such a large value of z would occur extremely rarely by chance, so we conclude that the variables (and their composite, the discriminant function) can be used to discriminate the two groups very effectively. The normal test is only a rough guide, however, since it takes no account of the number of variables involved, and this affects the outcome of the test. A more exact test is given by:

$$F = n_1 n_2 (n_1 + n_2 - k - 1) V(Y) / m (n_1 + n_2)(n_1 + n_2 - 2),$$

where m is the number of variables. This value can be looked up in F tables for m and $(n_1 + n_2 - m - 1)$ degrees of freedom. In our example,

$$F = \frac{80.60\ (80 + 60 - 3 - 1)}{3(80 + 60)(80 + 60 - 2)}\ 5.156 = 58.072$$

for 3 and 136 d.f. This value of F is highly significant.

Are all the variables equally useful for discriminating between the groups? If the standard deviations of the variables were about the same, then one could see which were the important variables by looking at the size of their corresponding weights. Clearly, if one variable has a small weight relative to the others, then it will not contribute much to the value of Y. When the variables have different standard deviations, however, we must first multiply each weight by the standard deviation of the corresponding variable before comparing the relative influence of each variable on Y. The square roots of the diagonal entries in V give us the standard deviations of the three variables. Hence:

$$\sigma(X_1) = \sqrt{14} = 3.742, \qquad \sigma(X_2) = 2.00, \qquad \sigma(X_3) = 4.472.$$

Thus,

$$w_1 \delta(X_1) = 0.165, \qquad w_2 \delta(X_2) = 0.034,$$
$$w_3 \delta(X_3) = 0.921.$$

From these figures it is apparent that father's income is the most important determinant of Y, whilst years of education is the least. Hence one can eliminate variables which do not appear important in discriminating groups after an examination of the weights. However, one must be careful in interpreting the meaning of the weights. Consider the covariance matrix below:

$$V = \begin{array}{c} \begin{matrix} X_1 & X_2 & X_3 \end{matrix} \\ \begin{bmatrix} 1 & 0.90 & 0 \\ & 1 & 0 \\ & & 1 \end{bmatrix} \end{array}.$$

All the variables have equal variances (all the diagonal entries are equal to unity), and variables 1 and 2 are highly correlated (their covariance is 0.9). If the values of

d_i are as in our example, then the weights that emerge turn out to be $w_1 = 11.578$, $w_2 = -8.422$ and $w_3 = 10$. The weight for the second variable thus turns out to be negative. It would appear, therefore, that as variable 2 increases, Y decreases, i.e. the greater the number of years of education, the more likely is that person to be classified as a Democrat supporter.

This appears to contradict the original data, in which the mean of variable 2 is greater for the Republican group. The explanation for this apparent paradox is that we cannot look at the weights and the variables in isolation. A high correlation exists between variables 1 and 2, so that a person who gets a high score on variable 2 (tending to decrease Y) is also likely to get a high score on variable 1 (tending to increase Y); the net effect may therefore be to increase Y. We can only really look at the weights and variables in relative isolation when the correlations between variables is not high.

7.6 Classification on the basis of discriminant function scores

If we have no reason to believe that the number of people in the population who are Democrat supporters is any different from those who are Republican supporters, then we can choose a point, Y_c, on the discriminant function, halfway between $\bar{Y}_1$ and $\bar{Y}_2$, and say that any person whose Y score falls below Y_c will be classified as Democrat and any person for whom $Y > Y_c$ will be classified as Republican (see Fig. 7.1). In our example, $Y_c = 7.78$. Referring to Fig. 7.1, the shaded portion shows the proportion of Democrat supporters who would be misclassified by this criterion. This proportion may be calculated by converting Y_c to its corresponding unit normal deviate and looking up the normal probability tables. Since $\sigma = 1$, $z(Y_c) = 7.78 - 6.64 = 1.14$. Hence the shaded portion represents a probability of misclassification of Democrat supporters of 0.1271 or 12.71 per cent. Because Y_c is as far from $\bar{Y}_1$ as $\bar{Y}_2$, the proportion of misclassified Republican supporters is also 12.71 per cent, and hence the total misclassification rate is 25.42 per cent. Placing Y_c midway between $\bar{Y}_1$ and $\bar{Y}_2$ is the optimum position for the cut-off point; moving Y_c to the right or the left increases the total misclassification rate. Suppose we move Y_c towards the Republican group mean by 0.1; the proportion of Democrat supporters misclassified decreases to 10.75 per cent, whilst the proportion of Republican supporters misclassified increases to 14.92 per cent. Hence the total misclassification rate increases from 25.42 to 25.67 per cent.

Another way of representing the error rate is by means of a contingency table. This is presented because it is the most convenient form when more than two groups are involved. The discriminant function score for each person in each sample is obtained and the number of people classified as Democrat or Republican correctly or incorrectly is calculated.

In our example, the results are shown in Table 7.2.

TABLE 7.2 *Classification of Democrats and Republicans on basis of discriminant function*

		Predicted group	
		Democrat	Republican
Actual group	Democrat	52	10
	Republican	8	70

Hence, 8 Democrat supporters and 10 Republicans are misclassified by the discriminant function.

Just as we can calculate the probability of misclassification for groups as a whole, so we can perform the same calculations for individuals. Suppose a person obtains scores on the three variables $X_1 = 20$, $X_2 = 8$ and $X_3 = 30$, so that his discriminant function score, Y, is 7.196; since this is less than Y_c, we classify this person as Democrat. What is the probability that this person is actually Republican? We look up the probability associated with a deviate from the Republican mean as large as $z = 7.196 - 8.91 = -1.714$. This probability is 0.0436, so that we can be reasonably confident about the correctness of our classification for this person.

7.7 The seriousness of misclassification

For some problems, the optimum position of the cut-off point, Y_c, is not midway between the means. Suppose, for example, the two groups represent diagnostic groups, 1 and 2, and people classified as 1 are to receive some risky treatment. Then it may be more serious to misclassify people as group 1 than group 2; thus we should move Y_c towards the mean of group 1. We can choose the position of Y_c such that the probability of misclassification of people as members of group 1 is made as small as we like. Again, using weights scaled in such a way that the within-group variances are unity, suppose $\overline{Y}_1 = 2$ and $\overline{Y}_2 = 5$. Suppose we are prepared to tolerate a probability of misclassifying members of group 2 equal to 0.01 or 1 per cent. Then we should choose Y_c such that $z = Y_c - 5$ cuts the distribution of Y for group 2 in such a way that the smaller proportion equals -2.33. The value of z required is -2.33 and hence $Y_c = 2.67$. The probability of misclassification for group 1 is now equal to 0.2514.

7.8 Unequal *a priori* probabilities

Another situation in which the optimum cut-off point may not be midway between $\overline{Y}_1$ and $\overline{Y}_2$ occurs when we have reason to believe that the probability of a person belonging to one group differs from that of him belonging to the other. Suppose, on the basis of recent general election results (to use our example), we believe that the ratio of Democrat to Republican supporters is 6:4. Suppose we choose a person at random; what is the probability that he will be misclassified if we choose Y_c to be midway between $\overline{Y}_1$ and $\overline{Y}_2$? A misclassification can arise in two ways: firstly, the person could be Democrat (with probability 0.6) and classified Republican (with probability 0.127), giving an error probability of $0.6 \times 0.126 = 0.0762$; secondly, a person could be Republican (with probability 0.4) and be misclassified with probability 0.127. This error can occur with probability $0.4 \times 0.127 = 0.0508$; hence the total error probability is $0.0762 + 0.0508 = 0.1270$. This error probability can be reduced by moving Y_c towards the mean of the group with the smallest *a priori* probability, i.e. $\overline{Y}_2$. The position of Y_c which gives the smallest value for the total error probability is the optimum cut-off point. This can be determined by trial and error, by moving Y_c in steps of 0.10 and determining the error probability, p. Let q_1 and q_2 be the *a priori* probabilities for groups 1 and 2, and p_1 and p_2 the smaller proportion of the discriminant function curves cut off by Y_c for groups 1 and 2 respectively. The trial and error procedure is summarised below; the first line of this table has already been calculated above.

TABLE 7.3 *Determination of optimal cut-off point unequal* a priori *probabilities*

Y_c	q_1p_1	q_2p_2	p
7.78	0.072	0.0508	0.1270
7.88	0.0645	0.0597	0.1242
7.98	0.0541	0.0694	0.1235
8.08	0.0449	0.0802	0.1251

There is no need to go beyond line 4 because the values of p at first decrease and then increase. The minimum value of p (0.1235) corresponds to a cut-off point of $Y_c = 7.98$. This value therefore minimises errors of classification for these particular *a priori* probabilities.

7.9 An alternative way of presenting the discriminant values

In this section we present an alternative procedure which is equivalent to the foregoing discriminant analysis, but has the advantage that it aids the process of classification when more than two groups are involved. This procedure gives a different set of weights for each group rather than a single set. Let us refer to these weights as g_{ij} where i stands for the variable and j the group. Then the weights are given by:

$$g_{ij} = \sum_{l=1}^{m} u_{il}\overline{X}_{lj} \tag{7.4}$$

where u_{il} are the elements of the inverse matrix U and, as before, there are m variables.

For example, for group 1 ($x = 1$),

$$\begin{aligned} g_{11} &= u_{11}\bar{X}_{11} + u_{12}\bar{X}_{12} + u_{13}\bar{X}_{13} \\ &= (0.102 \times 34) + (-0.094 \times 12) + (-0.012 \times 35) \\ &= 1.920. \end{aligned}$$

Similarly, $g_{21} = 0.088$ and $g_{31} = 1.238$.
In addition, we have to calculate a constant c_j, given by:

$$c_j = (\sum_{l=1}^{m} \bar{X}_{lj} g_{lj})/2. \tag{7.5}$$

$$\begin{aligned} \text{For group 1,} \quad c_1 &= \tfrac{1}{2}(\bar{X}_{11}g_{11} + \bar{X}_{21}g_{21} + \bar{X}_{31}g_{31}) \\ &= \tfrac{1}{2}(34 \times 1.92 + 12 \times 0.088 + 35 \times 1.238) \\ &= 54.833 \end{aligned}$$

For group 2, using equations (7.4) and (7.5), we obtain:

$$g_{12} = 1.82, \qquad g_{22} = 0.05, \qquad g_{32} = 0.770, \qquad c_2 = 37.175.$$

The relationship between the original weights, w_i, and these new weights, g_i, is given below. It will be useful to refer to the original weights as ${}_{12}w_i$, where the prefix makes it clear that they correspond to the discriminant function that best differentiates between groups 1 and 2. Then we have:

$${}_{12}w_i = g_{i1} - g_{i2} \tag{7.6}$$

In order to decide into which group to assign a person whose scores on X_1, X_2 and X_3 are known, we use the formula:

$$Y_j = X_1 g_{1j} + X_2 g_{2j} + X_3 g_{3j} - c_j \tag{7.7}$$

There are as many values of Y_j as there are groups; the highest value tells us which group that person is most likely to belong to. In our example, suppose a person obtains $X_1 = 20$, $X_2 = 8$ and $X_3 = 30$. Then we have:

$$\begin{aligned} Y_1 &= (20 \times 1.92) + (8 \times 0.088) + (30 \times 1.238) - 54.833 \\ &= 21.411, \end{aligned}$$

and

$$\begin{aligned} Y_2 &= (20 \times 1.82) + (8 \times 0.05) + (30 \times 0.770) - 37.175 \\ &= 22.725 \end{aligned}$$

Since $Y_2 > Y_1$, we would classify this person as being a Democrat supporter.

The classification criterion of using the highest value of Y_j is precisely the same as deciding on the basis of whether the persons discriminant function score falls above or below Y_c, where Y_c lies midway between $\bar{Y}_1$ and $\bar{Y}_2$. The two methods produce identical results, but the second is more convenient when more than two groups are involved.

7.10 Generalisation to more than two groups

We will now generalise the above procedure to take into account the case where there are more than two groups using a pair-wise comparison. In other words, we have as many discriminant functions as we have pairs of groups, three functions with three groups, six with four groups, and so on. In a later section (canonical discriminant analysis) we will consider a method which, in general, produces fewer discriminant functions than this.

We will illustrate multiple-group discriminant analysis with an example drawn from the field of labour economics which deals with the labour supply of married women. This example is based on an abridged version of work done by Gramm [1973]; a more detailed reference is made to this paper at the end of this section.

The following hypothetical data, shown in Table 7.4, refers to the mean scores of three groups of married women, differentiated as to their employment status, on three variables: X_1, husband's weekly wage; X_2, wife's weekly wage (actual or potential); X_3, average age of children.

TABLE 7.4 *Mean scores of married women on three variables, by employment status*

Variable	Group 1, full-time employed	Group 2, part-time employed	Group 3, not employed
Husband's wage, X_1	25.0	30.0	37.8
Wife's wage, X_2	22.0	20.0	18.8
Average age of children, X_3	9.0	6.0	4.2
Number in group, n_i	60	80	100

Just as in the two-group case we compute the within-groups sums of squares and cross-products matrices for each of the three groups, S_1, S_2 and S_3, and add them to obtain the pooled within-groups sums of squares and cross-products matrix W. Once again we have to assume multi-normal distributions and equal variances and covariances in each group.

We divide W by the appropriate number of degrees of freedom,

$$\sum_{i=1}^{k} n_i - k$$

(where k is the number of groups), to obtain the within-group variance-covariance matrix V, shown below:

$$V = \begin{bmatrix} 5 & 2 & 1 \\ 2 & 3 & 0 \\ 1 & 0 & 2 \end{bmatrix}.$$

The inverse of V, using the square-root method, is,

$$U = \begin{bmatrix} 0.315 & -0.210 & -0.158 \\ -0.210 & 0.473 & 0.105 \\ -0.158 & 0.105 & 0.579 \end{bmatrix}.$$

We can now use U to obtain the weights g_{ij} and the constant term c_j by applying equations (7.4) and (7.5), and these are shown below:

	Weights g_{ij}		
	Group 1	Group 2	Group 3
X_1	1.833	4.302	7.295
X_2	6.101	3.790	1.395
X_3	3.571	0.834	-1.567
c_j	106.093	104.932	147.698

7.11 Classification of individuals

As in the two-group case we can use equation (7.7) to determine to which group a particular woman is most likely to belong on the basis of her score on each of the three variables. Suppose, for example, a woman obtains scores of $X_1 = 35, X_2 = 17$ and $X_3 = 5$, then application of (7.7), using the weights in the above table, yields the following Y_j values:

$$Y_1 = 79.634, \qquad Y_2 = 114.238, \qquad Y_3 = 123.507.$$

Since the highest value is Y_3, we would classify this woman as a likely member of group 3, the 'not employed'.

To obtain an idea of how successful the discriminant analysis has been we can set up a contingency table, as in the two-group case, by calculating Y_j for each individual and assigning that individual to the most probable group on the basis of the highest Y_j score. The following contingency table, Table 7.5, has been deliberately constructed (and is not related to the data in the current example) in order to bring out some points of interest.

TABLE 7.5 *Contingency table for classification*

		Predicted group			
		Group 1, full-time employed	Group 2, part-time employed	Group 3, not employed	Total
Observed group	Group 1	50	0	10	60
	Group 2	0	65	15	80
	Group 3	10	40	15	100

If samples are reasonably large, we can use the figures in the above contingency table to estimate the probability of misclassification. Analytical methods exist for determining these probabilities but with large samples the contingency-table approach is adequate. The first point of interest is that whereas the probabilities of misclassification for women in full-time and part-time employment (groups 1 and 2) are fairly low, 10/60 = 0.167 and 15/80 = 0.188 respectively, that for the 'not employed' group is much higher, viz. 50/100 = 0.500. Consideration of these probabilities shows that we are not able to discriminate the 'not employed' group very effectively with the variables used here, since the probability of error is so high. Thus, in these circumstances, the researcher might employ other variables which could improve the situation: household assets, number of children, age of wife, etc. might be possible candidates for inclusion in the study.

The fact that group 3 is discriminated inadequately by the existing variable set also contaminates the discrimination between the other two groups. It can be seen from the contingency table that none of the group 1 individuals are misclassified as belonging to group 2 or vice versa. It appears, therefore, that these two groups can be reliably differentiated and the misclassification rates of these arises from a failure to properly discriminate between them and group 3.

One way of improving the situation if extra variables are unobtainable might be to amalgamate groups 2 and 3 into a single category, 'part-time employed and unemployed'; the probability of misclassification may thus be decreased. Alternatively, the unemployed group could be dropped completely from the analysis and the misclassification rate between groups 1 and 2 (the full and part-time employed) would drop to nearly zero. However, if this is done, a problem arises which might indeed arise in any analysis; that is, it must be assumed that a person belongs to one of the groups in the analysis, since the discriminant functions will assign the person to one of the groups even though he or she may not belong to any. For instance, in the two-group example considered earlier, an individual who is apolitical would nonetheless be assigned either as a Democrat or a Republican, which would be an obvious inadequacy in the design of the experiment. To illustrate this point, suppose, using the political example again, a person obtained scores on the variables as follows, $X_1 = 36$, $X_2 = 14$ and $X_3 = 20$. Initially, we can calculate this person's score on the discriminant function by applying equation (7.1), using the weights scaled for unit within-group variances, i.e. $Y = 5.94$. Since this value is less than $Y_c = 7.78$, we would classify the individual as a Democrat. However, we can calculate the chance that he belongs to the Democrat group as follows. Firstly, we calculate values for h_i, which is the difference between the mean for the Democrat group as a whole on variable i and that person's score on the same variable, i.e. $h_i = \bar{X}_i - X_i$. Thus we can determine $h_1 = 30 - 36 = 6$, $h_2 = -4$ and $h_3 = 5$. Next we obtain values b_i from:

$$b_i = \sum_{j=1}^{m} u_{ij} h_j,$$

e.g. $b_1 = u_{11}h_1 + u_{12}h_2 + u_{13}h_3 = (0.102 \times -6) + (-0.094 \times -4) + (-0.012 \times 5)$;

therefore

$$b_1 = -0.29.$$

Similarly,

$$b_2 = -1.09, \quad b_3 = 0.48.$$

Finally, we obtain $\Sigma h_i b_i$ which is distributed as χ^2 with m degrees of freedom.
In our example,

$$\chi^2 = (-6 \times 0.29) + (-4 \times -1.09) + (5 \times 0.48) = 8.50.$$

This value of χ^2 with 3 d.f. occurs with a probability less than 0.05. Hence it is unlikely that this hypothetical person is in fact a member of the group of Democrat supporters. A similar procedure can be carried out to ascertain the chance that the person belongs to the Republican group; this calculation produces a χ^2 of 16.61 with 3 d.f. Since this value occurs less than 1 per cent of the time by chance, it appears that a person with this pattern of scores is unlikely to be a member of either group although the discriminant analysis assigns him to one. It is, therefore, important to be sure that a person is a member of one of the groups in the analysis if one wishes to base his classification on the discriminant analysis of those groups, and it is a hazardous procedure to drop a group from the analysis if one wants to use the discriminant function for prediction purposes. Of course, if all one is interested in is seeing whether groups can be discriminated by means of the variables under study, this difficulty does not arise.

7.12 Canonical discriminant analysis

The foregoing procedure is not very parsimonious. It produces a set of discriminant functions each of which is a weighted composite of scores on the m variables that best separates two groups; hence there is one function for each pair of groups. If there are k groups, there are therefore ${}^kC_2$ discriminant functions.

In canonical discriminant analysis, the discriminant functions are called canonical variates. The first canonical variate is that weighted composite of scores on the m variables which maximally discriminates between *all* the groups. It may distinguish well between some of the groups but not others, in which case the canonical analysis will enable us to obtain a second canonical variate which is again a weighted composite which best separates the groups, with the proviso that it is orthogonal or uncorrelated with the first. Further canonical variates have similar properties, i.e. providing maximal separation between the groups whilst being orthogonal to previous discriminant functions. This discriminant analysis produces fewer discriminant functions than the previous method, and the maximum number required is equal to m or $(k - 1)$, whichever is the smaller, although when m and k are fairly large, considerably fewer than this are normally needed.

It is rather difficult to describe the computational steps for a canonical analysis in the simple terms needed for an introductory text such as this, so we will not attempt to do so; in any case, computer programs are readily available. Instead, we will merely present the results of a canonical analysis conducted on the data of the previous example on married female labour supply and interpret them.

The canonical discriminant analysis yields values for the coefficients (or weights) w_{ij}, for each variable i on each discriminant function j, for each canonical variate. These weights, produced by a computer program, are shown in Table 7.6 below. The values, λ_j, in the table represent the amount of discriminable variance associated with each discriminant function: it will be observed that the first function accounts for a large proportion of the total, which means that this first discriminant function acts highly efficiently in separating the three groups.

TABLE 7.6 *Canonical discriminant function weights*

	Discriminant function weights, w_{ij}	
	First discriminant function	Second discriminant function
Husband's wage, X_1	0.3036	– 0.0028
Wife's wage, X_2	– 0.1456	– 0.0225
Average age of children, X_3	– 0.1682	– 0.0257
Discriminable variance, λ_i	3021	25

7.13 Significance of the above discriminant functions

We first determine whether all the discriminant functions (in this example, two) significantly discriminate between the groups. This is done by calculating $\Sigma\lambda_i$, which is distributed as χ^2_n (where n refers to the number of discriminant functions) with $m(k-1)$ d.f. In our example, $\chi^2_n = 3046$ with 6 d.f. This significant value tells us that at least one discriminant function is significant; and if there is one which is significant, it must be the first because this accounts for the largest proportion of the discriminant variance. We next extract the amount of discriminable variance attributable to the first discriminant function from the total variance and the remainder is distributed as χ^2_{n-1} with $m(k-1) - (m+k-2)$ d.f. Here, $(m+k-2)$ is the number of d.f. associated with the first discriminant function, which has to be subtracted from the total d.f., $m(k-1)$. In our example, $\chi^2_{n-1} = 25$ with 3 d.f., which again is significant. Since there are only two discriminant functions altogether, this means that the second also discriminates between the groups. It was pointed out earlier that with three variables and three groups only two discriminant functions are possible. If, however, we had a larger number of variables and/or groups, then a significant value for χ^2_{n-1} would mean that at least one discriminant function other than the first was significant and this would in turn mean that at least the second was significant. We would then continue the process by subtracting the amount of discriminable variance attributable to the second discriminant function associated with which is $(m + k - 4)$ degrees of freedom and then test χ^2_{n-2} for significance with $m(k - 1) - (m + k - 2) - (m + k - 4)$ d.f. Subsequent discriminant functions have $(m + k - 6)$, $(m + k - 8)$ d.f. and so on.

7.14 Mean discriminant function scores

We obtain group means on the two canonical variables in a manner analagous to the application of equation (7.2). The mean score on the ith discriminant function of the jth group, $\overline{Y}_{ij}$, is given by,

$$\overline{Y}_{ij} = \sum_{l=1}^{m} w_{il} \overline{X}_{lj},$$

where the w and $\overline{X}$ values are taken from Tables 7.4 and 7.6 respectively. For example, the mean score for the first group on the first discriminant function is given by:

$$\overline{Y}_{11} = w_{11}\overline{X}_{11} + w_{21}\overline{X}_{21} + w_{31}\overline{X}_{31},$$

i.e.

$$\overline{Y}_{11} = (0.3036 \times 25.0) + (-0.1456 \times 22.0) + (-0.1682 \times 9.0) = 2.873.$$

This is the average value on the first discriminant function for women in full-time employment. The remaining mean values are shown in Table 7.7 below.

TABLE 7.7 *Mean discriminant function scores,* $\overline{Y}_{ij}$

	Group 1, full-time employed	Group 2, part-time employed	Group 3, not employed
First discriminant function mean	2.873	5.187	8.032
Second discriminant function mean	– 0.796	– 0.688	– 0.637

These results can be displayed graphically by plotting the group means with respect to the two axes representing the two discriminant functions and are shown in Fig. 7.2 below. These axes are orthogonal since the analysis ensures that successive canonical variates are orthogonal, i.e. are uncorrelated.

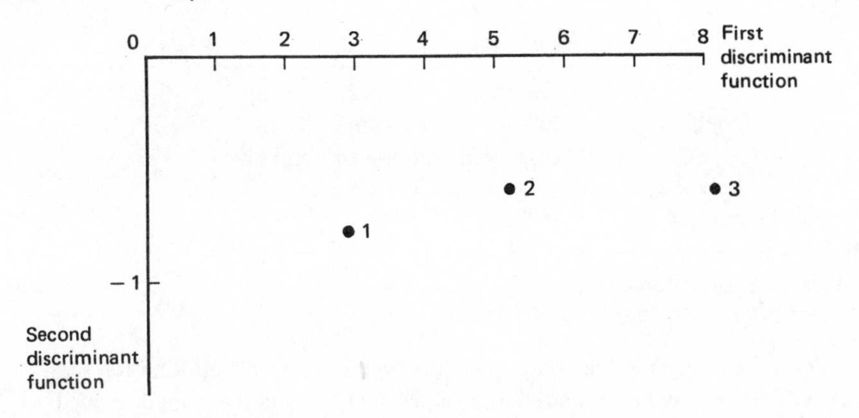

Fig. 7.2 *Graphical display of discriminant function means*

It is clear from the diagram that the three groups are well separated on the first discriminant function, and whilst the second function is statistically significant its discriminating power is not so obvious since the three are hardly separated at all. To all intents and purposes, we could regard the three groups as being discriminated on a single dimension.

Can we interpret this dimension in any meaningful way? We can multiply each discriminant weight, w_i, by the standard deviations of each of the variables to see which variable(s) contributes most to the discriminant function. The standard deviations are given by the root of the entries on the principal diagonal of the variance - covariance matrix, V. These are $\sigma_1 = 2.236$, $\sigma_2 = 1.732$ and $\sigma_3 = 1.414$; thus $w_1\sigma_1 = 0.679$, $w_2\sigma_2 = -0.252$ and $w_3\sigma_3 = -0.238$. It can be seen that the most influential variable is X_1, i.e. husband's wage. The positive sign of $w_1\sigma_1$ indicates that as husband's wage increases for some person so does the value of the discriminant function for that person. Conversely, the negative signs of $w_2\sigma_2$ and $w_3\sigma_3$ indicate that as wife's earnings and age of children increase, the value of the discriminant function decreases. However, as all three variables seem to contribute significantly to the discriminant function it is difficult to ascribe any meaningful interpretation to it.

To bring out some of the more interesting points about interpretation of canonical discriminant functions, we present an example using the same variables and groups as before but constructed in such a way that the means and within-groups variances and covariances have been altered so as to produce a solution having two discriminant functions which discriminate between the groups and in which both make a substantial contribution to the variance. We hope that the subsequent discussion as to the possible interpretation of the two functions will prove helpful. The means and canonical discriminant weights are shown in Table 7.8.

TABLE 7.8 *Means and canonical discriminant weights*

Variable	Group 1, full-time employed	Group 2, part-time employed	Group 3, not employed	Canon discrim wts. w_1	w_2
Husband's wage, X_1	25	27	32	0.753	– 0.052
Wife's wage, X_2	22	20	18	– 0.508	0.118
Average age of children, X_3	9	4	5	– 0.121	0.906

From the weights in the above table, the average discriminant function scores for each group may be computed using equation (7.2) and these are displayed graphically in Figure 7.3 below.

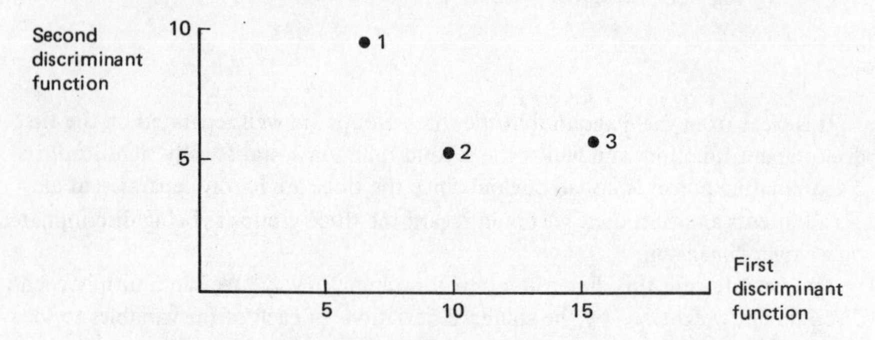

Fig. 7.3 *Graphical display of mean discriminant scores*

Can we meaningfully interpret these discriminant functions on the basis of the original variable set? The diagram suggests that whilst all three groups are widely separated in the first discriminant function, groups 2 and 3 are hardly separated at all on the second, whilst the first group is clearly distinguished.

On the first discriminant function, variables 1 and 2 have relatively high weights (0.753 and – 0.508), whilst variable 3 has a low weight (– 0.121). Since both variables 1 and 2 are concerned with income, we may refer to this discriminant function as describing a general income dimension. Thus, as husband's wage decreases or wife's earnings increase, the woman is more likely to be employed, the extent of this employment increasing as the net value of this general income dimension decreases. The age of children variable, X_3, plays no significant role in this dimension. However, in the case of the second function, both X_1 and X_2 appear to be relatively unimportant, whilst X_3 is. Hence, we might interpret this second function as being concerned with the average age of the women's children; the higher the average, the more likely it is that the women will be in full-time employment.

Since groups 2 and 3 are not discriminated well on this second dimension, we can say that consideration of income is what distinguishes between women in part-time employment and women not in employment, whilst fully-employed women are distinguished from both the other groups on the children's age variable in addition to income considerations.

7.15 Classification of individuals, the canonical case

We dealt with classification of individuals under the normal discriminant analysis procedure in a previous section in the example of female labour supply. In that case we used the individual's score on each of the three discriminant functions produced by the method. This resulted in a classification of a woman with scores of $X_1 = 35$, $X_2 = 17$ and $X_3 = 5$ into the 'not-employed' group.

We now want to deal with classification in the canonical discriminant function case. Our canonical analysis produced two canonical variates (or discriminant functions), one of which accounted for most of the variance (a considerable increase in parsimony); we will show that classification of the individual concerned can be achieved on the basis of this first canonical variate. The procedure is as follows:

The first step is to set up the vectors f_j whose elements consist of the canonical variate mean scores for each discriminate function for each group j, i.e. the $\overline{Y}_{ij}$ values of Table 7.7.

Thus,

$$f_j = [\overline{Y}_{1j} \quad \overline{Y}_{2j} \quad \overline{Y}_{3j} \text{ etc.}].$$

In our example,

$$\begin{aligned} f_1 &= [2.873 \quad -0.796] \\ f_2 &= [5.187 \quad -0.688] \\ f_3 &= [8.032 \quad -0.637], \end{aligned}$$

Although, since we are only concerned with the first canonical discriminant function, these reduce to:

$$f_1 = 2.873, \qquad f_2 = 5.187, \qquad f_3 = 8.032$$

(for simplicity we will refer to the elements y_{ij}, not to the vectors f).

We next transform the original raw score of the individual in question on the three variables, $X_1 = 35$, $X_2 = 17$ and $X_3 = 5$, into canonical scores by applying the canonical discriminant function weights w_{ij} of Table 7.6, and using the expression,

$$z_j = \sum_{i=1}^{m} w_{ij} X_i,$$

for the canonical score on the jth discriminant function.

In our example, for the first discriminant function,

$$z_1 = w_{11}X_1 + w_{21}X_2 + w_{31}X_3$$
$$= (0.3036 \times 35) + (-0.1456 \times 17) + (-0.0257 \times 5),$$

i.e.

$$z_1 = 7.3098;$$

similarly

$$z_2 = -0.609.$$

This enables us to form the canonical scores vector:

$$z = [7.3098 \quad -0.609].$$

We can now calculate the sum of squared deviations, D_k^2, of the individual's canonical variate scores about the mean canonical variates for each of the k groups, using the expression,

$$D_k^2 = \sum_{j=1}^{n} (z_j - \bar{y}_{jk})^2,$$

for the jth discriminant function, $j = 1, 2, \ldots n$.

Thus for the first group,

$$D_1^2 = \sum_j (z_j - \bar{y}_{j1})^2$$
$$= (z_1 - \bar{y}_{11})^2 + (z_2 - \bar{y}_{21}^2)^2 + \ldots + (z_n - \bar{y}_{n1})^2.$$

In our example, if both discriminant functions are used, we would have:

$$D_1^2 = (7.3098 - 2.873)^2 + (-0.609 - (-0.796))^2.$$

However, since we are concerned only with the first function, we can leave out the second term and we then have, for group 1,

$$D_1^2 = (7.398 - 2.873)^2 = 19.685.$$

Similarly,

$$D_2^2 = 4.506,$$

and

$$D_3^2 = 0.522.$$

We can now calculate the probability that a married woman with scores such as those above (i.e. $X_1 = 35$, $X_2 = 17$ and $X_3 = 5$) will belong to each of the three groups. We assume initially that each group in the population is of about the same size (this is the assumption of equal base rates). To calculate these probabilities we use the expression,

$$P(k/z) = e^{-0.5D_k^2} / \sum_i e^{-0.5D_i^2},$$

where the left-hand side is the probability of an individual with a canonical score vector of z belonging to group k. For the woman in question, the probability that she belongs to group 1 is thus:

$$\begin{aligned} P(1/z) &= e^{-0.5 \times 19.685} / (e^{-0.5 \times 19.685} + e^{-0.5 \times 4.506} + e^{-0.5 \times 0.522}) \\ &= 0.00006. \end{aligned}$$

Similarly, the probability of her belonging to group 2 is:

$$P(2/z) = 0.12006,$$

and to group 3 is

$$P(3/z) = 0.87990.$$

Thus we would assign this woman to the 'not-employed' group.

7.16 Unequal base rates

What difference do unequal population group sizes make? Suppose that the relative sizes of the three groups, employed, partially employed and not employed, are in the ratio 2:5:3. If we let π_j be the group population size, we have $\pi_1 = 0.2$, $\pi_2 = 0.5$ and $\pi_3 = 0.3$. To calculate the probabilities of an individual with canonical score Y belonging to each group k in turn when base rates are unequal, we use the expression,

$$P(k/z) = \pi_k e^{-0.5D_k^2} / \sum \pi_i e^{-0.5D_i^2},$$

where the π_i's are the relative population sizes.

Applying this expression to our example gives:

$$\begin{aligned} P(1/z) &= 0.00004 \\ P(2/z) &= P(3/z) = 0.8147. \ 0.1853 \end{aligned}$$

The probability is still that the individual will belong to the not-employed group, but the analysis has decreased this probability somewhat and taken into account the large size of the second group in the population, increasing $P(2/z)$ quite considerably.

It might be that the researcher wishes to impose further constraints on the assignment probabilities based on differential penalties associated with misclassification. One obvious example from medicine would be the undesirability of a patient not suffering from some particular disease undergoing surgery. We would wish to impose high penalty 'costs' on the possibility of misclassification in these circumstances. In this case the expression,

$$P(k/z) = C_k \pi_k e^{-0.5 D_k^2} / \sum_i C_i \pi_i e^{-0.5 D_i^2},$$

where the C_i are 'costs' associated with misclassification for each group.

Finally, we may note that some authors have suggested that we can test the classification of our discriminant functions by using only a part of our total sample of observations to determine the functions and then applying these to the remaining unused data to validate them. The feasibility of this procedure would, of course, depend on the size of the original sample.

7.17 Applications

We would like to conclude by referring briefly to three applications of discriminant analysis to be found in the literature.

The first application is one referred to earlier and is due to Gramm [1973]. The study concerns the labour supply of married women teachers and divides women into three groups (full-time, part-time and not employed) for the purposes of the analysis. The survey covered over 400 married women teachers (employed and unemployed) who were measured on their response to 10 variables, e.g. wage of husband, full-time wage of wife, part-time wage of wife, household assets, ages of children, etc. The analysis showed that the most important factors affecting the desired labour force status of married women teachers were the wages of the husband, the wages of the wife and the age of the first child (or number of children under six). However, the most important result for Gramm is that it is possible to distinguish between full-time and part-time workers on the basis of the variables used, and the two groups should not therefore be lumped together in labour participation studies as is usual.

The second application we want to mention is by Bledsoe [1973], who used discriminant analysis to predict teacher competence. Six groups of teachers were chosen (covering a total sample of 415) representing five various teaching fields, e.g. maths, science, English and a general teaching field. The teachers were divided into two groups depending on whether they scored 'low' or 'high' on a set of selected teaching competence tests based on COR (Classroom Observation Record). Each of the teachers was measured on 25 predictor variables: 20 derived from teacher self-reports from four tests, e.g. Teacher Practices Questionnaire (TPQ), Kerlinger Education Scale VII (KES), etc.; 5 variables were obtained from pupils' responses in the Pupil Observation Report (POR), a 38-item questionnaire. The analysis showed that the two groups were successfully discriminated on the 25 variables. In terms of proportion misclassified, the best results were for the science and the maths teachers (94 per cent correctly assigned in each case), the worst for social studies and general teachers (76 and 75 per cent).

One interesting result of the study is that different tests best predict teaching competence in different fields. For the general teacher, the best test of competence was Ryans TCS (Teacher Characteristics Schedule), whilst classification of maths teachers was best achieved by POSR and KES. The analysis also suggested that a 'good' teacher in one field may not be good in another.

The final application is drawn from a paper by Dobson *et al.* [1974], in which he discusses the application of multidimensional scaling to public attitudes towards a new transport system (see Chapter 6). He concludes his investigation by performing a discriminant analysis on the basis of the six dimensions which emerge from his MDS study, e.g. general level of service (assurance of a seat, punctuality), flexibility of service (longer hours, more routes), lower fares, etc. Each subject was tested to see whether he or she belonged to the upper or lower quartiles in terms of how important they rated each dimension and these two quartiles of importance were used as a basis for a two-group discriminant analysis on each of the six dimensions, using as predictors scores on 10 socio-economic-demographic variables, e.g. age, sex, marital status, number of automobiles in family, number of drivers in family, income range, etc. For all dimensions except the second, the discriminant function was able to distinguish significantly between the two quartiles at the 0.01 level. Dobson goes on to discuss the interpretation of the discriminant weights on each variable for each dimension, e.g. for the first dimension, respondents from families with an excess of licensed drivers over available automobiles placed considerable importance on high levels of service, but married females with high educational levels placed more importance on the conglomerate of amenity, aesthetic and social interaction attributes.

This concludes the discussion on some applications of discriminant analysis. The technique is extremely versatile and many possibilities exist for its further use in all the fields of the social and behavioural sciences. Frequently, the analysis of conventional problems in terms of sub-groups of the original observations will yield insight into aspects of the problem not previously considered.

8 The Analysis of Qualitatitive Data

8.1 Introduction

In previous chapters we have dealt principally with two problems, that of factor-analysing relationships between correlated variables and that of discriminating between groups and classifying individuals or objects into categories. Throughout, it was assumed that the measurement of the variables under consideration was in *quantitative* terms. Furthermore, it was also assumed that the variables in question were normally distributed (at least approximately) and that the regression relationship was linear.

However, very often in the social and behavioural sciences (in some areas of study, more often than not) we may be interested in variables which can only be measured in *qualitative* terms, for example, religion, sex, marital status, social class, political allegiance, etc. Some of these variables, such as sex, provide binary or dichotomous results, whilst others, such as religion, may provide multiple or polyotomous data.

With quantitative data we can compute product-moment correlations and factor-analyse matrices of such correlations. However, with qualitative data we may only be able to compute rank correlations, tetrachoric correlations, bi-serial correlations and so on.* As far as the first problem above is concerned, matrices of such correlation coefficients may still be factor-analysed by any of the previously discussed methods. But one cannot be quite as confident as to the quality of the results and any conclusions drawn from them must be interpreted with caution. If the purpose of a particular factor analysis on qualitative data is simply to identify clusters of similar variables, then analysis of such matrices may be satisfactory, but the researcher should beware of drawing conclusions other than this because the assumption of normal distributions of the variables would be violated. Another difficulty associated with the use of correlation coefficients derived from variables which display only a few categories is that they are less

*For a discussion of the meaning of these terms see, for instance, Baggeley [1964].

reliable (or reproducible) and more prone to distortion. This is especially true when there are a few extreme cases, in other words when the distribution is highly skewed. For example, with a dichotomous variable, we can give a score of 1 to one level of the variable and 0 to the other, and obtain, in the extreme case, say ninety-five 1's and five 0's. Correlations between such a variable and others can be markedly altered by changes in just a few values (and are therefore prone to instability), and the maximum value of such a correlation can be considerably less than one if it is included in the same data set as variables with less extreme splits. Thus, its communality will be low and its factor loadings will be low. It is wiser to avoid qualitative data in a factor analysis where possible, but if unavoidable then it is better at least to avoid extremes of skewness in the distributions.*

As far as the second problem is concerned, i.e. that of discrimination and classification, methods applicable to qualitative data will be presented later. Firstly, however, we want to describe a method for the factor analysis of qualitative data in which the data itself is collected in a way which is particularly interesting and known as the *repertory grid technique*. Following this, we will consider the problem of discrimination and classification in the presence of qualitative data and present a technique known as *the pattern-probability model*. Finally, we will discuss a method which has elements of both the analysis of relationships between variables and of classification of individuals or objects; this is known as *latent structure analysis*.

8.2 Repertory grid – factor analysis with qualitative data

Repertory grid techniques were originally developed in the area of psychological research† and so far most of its applications have been in this field, but it is quite possible for the technique to be applied more generally as we will see later. Basically, the repertory grid technique amounts to a series of sortings of a number of *elements*, according to a number of criteria or *constructs*. The elements are usually, in applications in psychology, significant people in the life of the patient (or 'subject' in an investigation), such as wife or husband, parents, best friend, most disliked person and so on; the constructs are attributes such as honest - dishonest, warm - cold, etc. The patient or subject is required to generate the constructs according to a set of procedures (an example of which will shortly be described) and then to use the constructs to classify or sort the elements. Following this procedure, a grid can be set up and factor analysed to see how the patient's constructs are related to one another. In this way the investigator can establish an insight into the patient's construct system, for example its degree of complexity. If only a few important factors emerge, this may indicate a com-

*Because rank correlations, tetrachoric correlations etc. are not as stable as product-moment correlations, the sample sizes required to produce correlations displaying the same stability need to be two or three times as large.

† The idea originated with Kelly [1955]. Applications and later developments are summarised in Bannister and Mair [1968].

paratively simple system reflecting a naive and over-generalised view of the patient's social relationships. Moveover, by obtaining a second grid sort at some later date from the same patient, perhaps following a period of therapy, and factor analysing this, the investigator can determine what changes, if any, have taken place in the complexity or other properties of the subject's construct system.

As indicated, the technique can be widely generalised into other areas of the social and behavioural sciences. For example, elements could be products, TV programmes, types of employment, film stars, religions, nationalities, political situations, etc., whilst the constructs would be chosen to correspond appropriately to the elements and to the investigator's particular interest. Thus for products, the constructs might be convenience, quality, appearance, reliability, price, durability, etc. Furthermore, there is no reason why individual grids only should be analysed; those obtained from groups of individuals can be analysed also. The difficulty here, however, is that unless there is a reasonable degree of agreement amongst the individuals as to how they interpret the elements, correlations between constructs will tend to be low and the researcher may be forced to abandon or modify this particular study. This problem may in fact arise in factor-analytic studies employing any type of variable measure, i.e. quantitative as well as qualitative, and the best safeguard in such circumstances is to choose a population as apparently homogeneous as possible and sample sizes as large as feasible. (In the field of marketing, market segmentation studies will naturally tend to produce homogeneous populations.)

One obvious application of the repertory grid technique is in the field of attitude surveys. Opinion researchers are frequently interested in the measurement of attitudes towards a variety of things. In the following example, which we have chosen to illustrate the technique, we are concerned with the measurement of attitudes to the personal characteristics of political figures. With a *conventional* attitude survey, the investigator would either construct open-ended questions designed to assess how his sample construes the objects of the investigation or, more probably, he might decide on the personal characteristics which he deems relevant and require his sample to assess the objects of the investigation with respect to those characteristics. The former method has the advantage that it does not put constraints on the subjects (i.e. they are free to employ constructs or descriptions which they, rather than the investigator, consider relevant or important) but this makes it harder to score and interpret than the second method. The *repertory grid* technique, however, combines advantages of both methods, as we shall see.

8.3 Public attitudes to politicians, the use of repertory grid: production of the constructs

In this fictitious study of public attitudes towards 10 prominent political persons (who form the *elements* in this example), the first step is to determine what it is about these politicians, what attributes or *constructs* are used by the *subject* (a member of the public) to form judgements about the politicians in question. These constructs are the variables which subsequently will be the subject of the factor analysis. We will probably discover that a member of the public produces a large number of seemingly quite different and unrelated constructs which, though, are unlikely to be independent. The purpose of the factor analysis is to examine the interrelationships between these initial character assessments and thus to discover the underlying dimensionality by which the individual (probably unknowingly) judges the politicians. In addition to the inclusion in the analysis of variables (or constructs) which arise spontaneously from the judgements of the subject, the investigator can, in addition, include variables which may be of particular interest. For instance, in our example we suppose that the particular interest of the researcher is to discover what personal characteristics of each politician relate to his attractiveness in terms of potential voting support, and thus the attribute 'votable for' is included in the set of variables.

It is the function of the repertory grid technique to elicit the original set of constructs on which the factor analysis may then be applied. There are a variety of ways in which this can be done (see, for example, Bannister and Mair [1968]), but one convenient approach (especially since it has already been described in the section dealing with multidimensional scaling) is the method of triads. In this particular example the approach is as follows. A photograph of each of the 10 politicians is first of all presented to the subject. He is then re-presented with the photographs three at a time and asked to name a way in which two of the politicians are alike and dissimilar to the third, on the grounds of some personal characteristic, for example politicians A and B seem 'honest' whilst politician C is not judged to be honest. The personal characteristic (and the name given to it) used by the subject in his decision must arise spontaneously and is not suggested or imposed by the researcher (although, as mentioned above, some personal characteristics or variables may be introduced by the researcher in the second stage of the method). The subject is presented with as many triads as is required to produce a large enough set of constructs (how large this is will depend on the researcher). Let us suppose that such a procedure has been followed and the following set of constructs has arisen, Table 8.1. Construct 4, 'votable for', is added subsequently by the researcher (assuming that it does not arise spontaneously from the subject during the trial process).

TABLE 8.1 *List of 16 constructs (or variables) on personal characteristics*

Constructs	
1. honest - dishonest	9. physically attractive - unattractive
2. strong - weak	10. Inspires confidence - does not
3. able - incompetent	11. in touch with people - aloof
4. votable for - not votable for	12. has vision - no vision
5. compassionate - insensitive	13. experienced - inexperienced
6. intelligent - stupid	14. determined - weak-willed
7. common sense - no common sense	15. friendly - unfriendly
8. good speaker - poor speaker	16. broad-minded - narrow-minded

8.4 Measurements of constructs

The second step in the repertory grid technique, following the determination of the set of constructs, is to measure each of the elements (in this case the politicians) against each of the constructs. There are several ways in which this may be done. One might be simply to ask the subject to assert whether each politician 'possesses' or 'does not possess' each construct in Table 8.1. A second possibility would be for the subject to rank each politician in order of preference, from one to ten, against each construct. A further possibility might be for each element to be ranked on a five-point scale as to the degree of his possession of each construct. When the sample of subjects is reasonably large and the results are to be aggregated, the choice of technique is not crucial.* In our example, the first method, 'possesses',

TABLE 8.2 *Scores for each politician by a particular subject on the 'possess', 'does not possess' criterion for each construct*

	Elements (politicians)									
Constructs (personal attributes)	1	2	3	4	5	6	7	8	9	10
1. Honest	1	0	1	1	0	0	0	1	1	0
2. Strong	0	0	1	0	1	1	0	1	1	1
3. Able	1	1	1	0	0	1	0	1	1	1
4. Votable for	1	0	0	1	0	1	0	0	0	0
5. Compassionate	0	0	1	0	1	0	1	0	0	0
6. Intelligent	0	1	0	0	0	1	0	1	1	0
7. Common sense	1	1	1	1	0	0	1	1	0	1
8. Good speaker	1	0	1	1	0	1	0	1	1	1
9. Physically attractive	0	1	1	1	1	1	0	1	0	0
10. Inspires confidence	0	0	1	0	1	1	0	1	1	1
11. In touch with people	1	1	1	1	0	0	1	1	1	1
12. Has vision	0	0	0	0	1	0	1	0	0	0
13. Experienced	1	1	0	0	1	1	1	1	0	0
14. Determined	0	1	1	0	0	1	0	1	1	1
15. Friendly	0	1	1	1	0	1	0	1	1	0
16. Broad minded	1	0	1	0	0	1	1	1	0	1

*Grid techniques are frequently employed in the study of individuals and in this case the use of the first method 'possesses', 'does not possess' criterion may lead to tetrachoric correlations with the attendant difficulties for subsequent factor analysis mentioned previously.

'does not possess', is used and a possible result for a particular subject is shown in the grid of Table 8.2 ('1' indicates 'possesses').

Since it is possible for extreme responses to cause skewness (for example the subject may regard politician 'X' as honest and all the others dishonest, thus producing a distribution of one '1' and nine '0's), some constraints may have to be imposed on the subject's freedom to sort. For instance, in such a case as that above, he may be required to allocate half the politicians into a 'more honest' category and the remaining half into a 'less honest' category. If this seems an unreasonable procedure (to the subject or the researcher), then one of the other methods of measuring the constructs, e.g. ranking on a five-point scale, may be employed.

From the grid in Table 8.2, correlations between each pair of attributes can be computed and the resulting correlation matrix can be factor analysed by any of the methods previously described. In this example, since variables are dichotomous, a tetrachoric correlation might be used. The kind of result which might be expected to emerge from the factor analysis is shown in Table 8.3, which shows the oblique factor loadings of four factors on each of the constructs; the significant loadings are in italic. Table 8.4 shows the correlations between these four factors.

TABLE 8.3 *Oblique factor loadings of four factors on 16 constructs (personal attributes): significant loadings in italic*

	Factor			
Construct	1	2	3	4
1	*0.39*	0.13	0.16	0.18
2	0.23	*0.80*	0.26	0.26
3	*0.43*	0.08	0.26	0.00
4	*0.49*	0.16	0.12	0.18
5	0.05	0.05	*0.75*	0.25
6	0.14	0.22	0.25	*0.59*
7	0.11	0.15	0.00	*0.43*
8	0.07	0.18	0.28	*0.50*
9	0.09	0.12	*0.55*	0.19
10	*0.31*	0.23	0.25	0.05
11	0.25	0.19	*0.67*	0.01
12	0.11	0.15	0.11	*0.51*
13	0.10	*0.34*	0.07	0.08
14	0.16	*0.56*	0.09	0.08
15	0.17	0.03	*0.42*	0.04
16	0.16	0.13	0.09	*0.75*

TABLE 8.4 *Correlations between factors*

	Factor			
	1	2	3	4
1		0.3	0.2	0.2
2			0	0.2
3				0
4				

8.5 Interpretation of factors

From examination of Table 8.3 and the significant construct (variable) loading on each of the four factors (leaving aside for the moment construct 4, 'votable for'), we can tentatively interpret the factors and suggest the following names: factor 1 with significant loadings on constructs 1 (honest), 3 (able) and 10 (inspires confidence)) we label 'trustworthy'; similarly for the other three factors. We thus have, Table 8.5,

TABLE 8.5 *Interpretation of factors*

Factor	Significant construct	Interpretation
1	1, 3, 10	trustworthy
2	2, 13, 14	forceful
3	5, 9, 11, 15	personable
4	6, 7, 8, 12	intellectual

(Of course, as always, interpretation of factors leaves much room for uncertainty and may call for some ingenuity on the part of the researcher in finding a single word or phrase that adequately encompasses the collective implications of a set of variables.)

Thus we can see that the 15 original constructs (i.e. personal characteristics) have been reduced by the factor analysis to a set of only four basic constructs which underlie the original larger number. What this implies is that although this particular subject appeared to judge the politicians on the basis of a large number of variables describing personal characteristics, in practice his (or her) construct system was much simpler. The researcher is particularly interested (we suppose) in what personal characteristics make any politician attractive to voters; in other words he is interested in that factor (or factors) in which construct 4, 'votable for', has a significant loading and the interpretation or meaning which can be ascribed to that factor. In this example, as can be seen from Table 8.3, construct 4 has a significant loading on the first factor, 'trustworthy', and the researcher may conclude, therefore, that of the four underlying factors, the perceived trustworthiness of any politician is the personal characteristic most closely associated with voting support. In any promotional campaign then, he may emphasise this aspect of the candidate's persona. In this particular application of repertory grid, of course, the investigator will be more interested in the judgements of a large sample of potential voters regarding the personal characteristics of the politicians.

There are two possible approaches. Firstly, having used the method of triads to obtain a set of constructs from each individual in the sample, we could select from this total set a smaller set of those which occur most frequently. This set could then be presented to the subjects, each of whom would be asked to sort the politicians in one of the ways indicated above, from which a grid could be obtained for each subject in the sample. The values in the corresponding cells of each of these grids would then be added to obtain an aggregate grid for the whole sample, each cell having a possible value in the range 0 to n, where n is the sample size. Because of this aggregation, the problem of skewness in the distribution of scores in individual grids is not so important, the aggregation making less likely a departure from normalcy. From the aggregate grid, product-moment correlations between each pair of attributes (constructs) can be computed and the resulting correlation matrix factor analysed and interpreted as before.

A second possible approach would be to factor analyse the data resulting from each individual separately (as above), and compare the resulting factor solutions for similarities. One drawback to the first approach is the possibility of inhomogeneity in the population being sampled (although in a practical application of the technique this possibility might be minimised by careful sample design). Using the second approach, the degree of homogeneity in the population may be assessed directly by examination of the degree of similarity amongst the factors which emerge from the analysis for each individual. There may well be some factors which occur frequently amongst the individuals. For example, if two construct clusters, which we have identified previously as say 'trustworthy' and 'personable', are common to many of the subjects, this would indicate assessment of the politicians on the basis of these two personal characteristics and this information might be useful to the researcher.

A further point of some interest is whether any key variable (in this example 'votable for') falls in the same (or similar) cluster of variables for most subjects, thus indicating a degree of homogeneity in the sample. If neither of these possibilities occur, i.e. if there is evidence of heterogeneity, then the researcher can attempt to sub-divide the population into what are, hopefully, more homogenous groupings and then perhaps assess the difference between them. This in itself may reveal quite useful information. Suppose in this example, we sampled men and women separately; it may prove that the two sexes use different key factors in assessing the appeal of political personalities. A political campaign could then be designed with this fact in mind.

Just as the clinical psychologist may perform the repertory grid techniques on several occasions, i.e. before and after therapy, so the procedure above could be repeated, say progressively, during a political campaign to determine whether the complexity of political judgements changes or whether the 'votable for', variable shifts into other factors.

8.6 The pattern probability model – discriminating between groups with qualitative data

The aim of the pattern probability model is to classify individuals into groups on the basis of 'patterns' of responses to certain qualitative variables. For example, suppose in a study on health, we wish to classify individuals into several different groups according to the probability of them suffering some particular chronic illness. The classification is to be based on the responses to three variables, sex, age (44 or under, over 44) and type of employment (manual, non-manual), on which each individual in the study is to be measured. The response of a particular individual to the three variables, e.g. 'male, over 44, manual', constitutes the *pattern* and the range of values which each variable can assume constitute the *elements* of the pattern. In this example, all the variables are dichotomous (although this will not always be the case). On the first variable a response of 1 is used to indicate male and 0 female; on the second variable a response of 1 indicates 44 or under and 0 over 44; for the third 1 indicates non-manual and 0 manual. The response pattern above (male, over 44, manual) could thus be coded, using the above dichotomous values, as $X = (1\ 1\ 0)$, where we use X to denote the pattern. (X is thus a pattern *vector*, whose elements correspond to the responses on each variable.)

The example we have chosen to illustrate the pattern probability model is based on data taken from the General Household Survey [1973] dealing with the incidence of four chronic diseases in the population, mental and nervous, heart and circulatory, respiratory and diseases of the musculo-skeletal system, coded 1, 2, 3 and 4 respectively. These four diseases form the four *classes* into which we want to assign individuals. The first aim is to determine the probability that individuals who are characterised by a particular response pattern belong to a certain class, i.e. suffer from a given chronic illness. The second aim is to use these results to make predictions about the relationships between responses on variables and diseases. In other words, we want to determine conditional probabilities such as $P(j/X)$, where j indicates the class, i.e. type of disease, and X the pattern of responses on the variables. For instance, in the example which follows, we are able to show that when the response pattern is $X = (0\ 1\ 1)$, then $P(j/X) = 0.51, 0.10, 0.22$ and 0.17 for $j = 1$ to 4 respectively. Since the maximum of $P(j/X)$ is this case 0.51 (corresponding to $j = 1$), it may be concluded that individuals displaying this response pattern, female, aged 44 or under and non-manual, are more likely to suffer from a mental or nervous disease than any other chronic illness, should they become chronically sick.

Having outlined the principle of the method we now proceed to describe the procedure in detail. Two versions of the model will be presented. In the first, known as the *equal base rates model* (or sometimes the maximum likelihood model), it is assumed that the incidence in the population of each of the four diseases is the same. In the second model, the *unequal base rates model* (or the Bayesian conditional probability model), the actual incidence of the diseases in the population is taken into account.

8.7 Equal base rates model

If we have a large enough sample of individuals in each group or if the number of different patterns is relatively small, then the conditional probabilities can be determined by a direct frequency count. If the number of patterns is large, which it can be with only a relatively small number of variables (for example, five variables, each of which is trichotomous, leads to a number of different patterns equal to $3^5 = 243$), then the required sample size would have to be very large to establish reliable probability estimates for each pattern. This difficulty can be overcome if we are able to make the assumption that no association exists between the variables within each class, i.e. there is no relationship between being male and being over 44, or being a male and being a manual worker, etc. Then we can make use of the multiplication law defining the joint probability of several independent events.* In practice, slight deviations from this assumption of independence do not appear seriously to distort the conclusions based on the method; in any case, its appropriateness, like that of any multivariate model, should be evaluated by replication and independent investigations where possible. Provided this assumption can be met and that we know the values of $p(x_i/j)$ for $i = 1, \ldots, k$, then we can use the multiplication law to obtain terms $P(X/j)$ and then calculate $P(j/X)$ using the equation:

$$P(j/X) = \frac{P(X/j)}{P(X/1) + P(X/2) + \ldots + P(X/n)} \qquad (8.1)$$

The actual data used in the example is shown in Table 8.6 below and is based on information contained in the General Household Survey. The table shows the

---

*The multiplication law states that:

$$P(A \text{ and } B \text{ and } C \text{ and} \ldots \text{and } K) = P(A) \times P(B) \times P(C) \times \ldots P(K),$$

where $A, B, \ldots, K$ are independent events.

Applied to the probability pattern model we are assuming that, given $X = (x_1\ x_2 \ldots x_k)$, where x_i is the response value to the ith variable, then $P(X/j) = P(x_1/j) \times P(x_2/j) \ldots P(x_k/j)$, where $P(X/j)$ means the probability of obtaining a pattern X within a class of disease j.

incidence of chronic sickness for four selected diseases (rate per 1000 of the population), although some liberties have been taken with the data in order to simplify the calculations.

TABLE 8.6 *Incidence of four selected chronic diseases per 1000 of population (from General Household Survey)*

	Variable and coding response					
	Male	Female	44 or under	over 44	Non-manual	Manual
Disease class and type	1	0	1	0	1	0
1. Mental and nervous	19.3	33.1	16.6	65.1	27.7	50.2
2. Heart and circulatory	46.6	57.5	7.4	20.1	59.9	80.0
3. Respiratory	35.7	31.1	14.7	117.5	33.1	77.6
4. Musculo-skeletal	48.7	68.2	12.4	199.5	61.3	96.5

From the data in Table 8.6 the values of the conditional probabilities $P(x_j/j)$ can be calculated. For example, $P(x_1 = 1/1)$ represents the proportion of males ($x_1 = 1$) reporting mental and nervous diseases ($j = 1$), and in this case,

$$P(x_1 = 1/1) = 19.3/(19.3 + 33.1) = 0.37.$$

Table 8.7 shows the complete set of the $P(x_i/j)$ values obtained in this way.

TABLE 8.7 *Conditional probabilities, $P(x_i/j)$, for chronic disease incidence of Table 8.6*

	Variable and coding response					
	Male	Female	44 or under	over 44	Non-manual	Manual
Disease class and type	1	0	1	0	1	0
1. Mental and nervous	0.37	0.63	0.20	0.80	0.36	0.64
2. Heart and circulatory	0.45	0.55	0.04	0.96	0.42	0.58
3. Respiratory	0.53	0.47	0.11	0.89	0.36	0.64
4. Musculo-skeletal	0.41	0.59	0.06	0.94	0.41	0.59

Using the information contained in Table 8.7, we can proceed to calculate the $P(j/X)$ probabilities using the multiplicative law. As an illustration of how this is done, suppose the pattern of responses for an individual is $X = (0\ 1\ 1)$, then for disease class 1 (mental and nervous diseases), the probability of obtaining this response pattern is given by,

$$P(X/1) = P(x_1 = 0/1) \times P(x_2 = 1/1) \times P(x_3 = 1/1),$$

which from row 1 of Table 8.7 gives,

$$P(X/1) = 0.63 \times 0.2 \times 0.36 = 0.045.$$

Similarly, the probability of obtaining this response pattern for disease class 2 (heart and circulatory diseases) is equal to,

$$P(X/2) = 0.009$$

In the same way, $P(X/3) = 0.019$ and $P(X/4) = 0.015$.

We can now use equation (8.1) to calculate the conditional probabilities, $P(j/X)$; for example,

$$P(1/X) = \frac{P(X/1)}{P(X/1) + P(X/2) + P(X/3) + P(X/4)}$$

$$= 0.045/0.088$$

$$= 0.51$$

Similarly,

$$P(2/X) = 0.10$$
$$P(3/X) = 0.22$$
$$P(4/X) = 0.17$$

Thus, if an individual displays the response pattern given above, i.e. $X = (0\ 1\ 1)$, then it is most probable that she will suffer from sickness class 1 (mental and nervous diseases), since $P(1/X)$ has the highest value of the four $P(j/X)$ values. This assertion is likely to be correct 51 per cent of the time and incorrect 49 per cent of the time. This compares to a *chance* assignment of being correct 25 per cent of the time, assuming equal incidence of each of the four diseases. In the next section we will discuss the consequences when this assumption is not true.

The total set of possible pattern responses and the corresponding disease classification, together with the probability of correct assignment resulting from the application of equation (8.1) is given in Table 8.8.

TABLE 8.8 *Variable response pattern and corresponding chronic sickness classification, together with probability of correct assignment*

Response pattern, X	110	110	101	100	011	010	001	000
Disease classification	1	1	2	3	1	1	4	4
Probability	0.41	0.43	0.29	0.31	0.51	0.55	0.29	0.27

Some interesting comments may be made on the results. Firstly, notice that assignment to class 1 disease occurs whenever there is a one in the middle position, i.e. the subject is recorded as being aged 44 or under. This implies that if an individual who is aged 44 or under becomes chronically sick with one of these four types of illness, then it is most likely to be a mental or nervous illness. This is regardless of the response on the other two variables, i.e. regardless of whether the individual is male or female or a manual or non-manual worker. Secondly, a class 4 illness (musculo-skeletal) is indicated when the response pattern has zeros in the first two positions, which corresponds to female, over 44. Thus, if a woman aged over 44 suffers from one of these four diseases, it is most probable that it will be musculo-skeletal in nature. Finally, for males aged over 44 (i.e. one in the first position and zero in the second) the most probable illness depends on type of employment: manual workers are more likely to suffer respiratory diseases, whilst non-manual workers are more likely to suffer from heart and circulatory diseases.

Some comment on the classification probabilities might also be made. For individuals who are classified as likely to have a class 1 disease, the probability of a correct classification ranges from 0.41 in the case of response pattern $X =$ (1 1 1) to 0.55 in the case of pattern (0 1 0). In the case of the other diseases, the results mean that we can be less confident in our prediction, a range from 0.31 at best to 0.27 at worst is not a great improvement on chance. We conclude, therefore, that in this application, the pattern probability model is only useful in predicting those individuals who are likely to suffer mental and nervous diseases. Thus, in this example, age is the only important predictor variable in terms of its influence on the likelihood of developing this form of chronic sickness.

8.8 Unequal base rates model

In practice, it would not generally be the case that the incidence of each disease in the general population is the same (the assumption of the equal base rates model) and we must therefore modify our technique to take this fact into account. We do this using the *unequal base rates model.*

First of all we assume that we can determine the actual incidence of each of these diseases in the population. In this particular example, the information is readily available from another source,* otherwise it might be necessary to undertake a pilot sample of the population in question. The necessary data is shown in Table 8.9: the first column records the percentage of the population as a whole who suffer from each of the four diseases; the relative proportions, π_j, for the jth disease, corresponding to these incidences are shown in the second column (this is merely a conversion of column one to produce unit sum).

**Social Trends* [1972]

TABLE 8.9 *Actual percentage incidence of chronic sickness in population and relative proportions, π_j*

Disease class	% incidence	Relative proportion j
1. Mental and nervous	0.0249	0.16
2. Heart and circulatory	0.0466	0.30
3. Respiratory	0.0357	0.23
4. Musculo-skeletal	0.0487	0.31

When *a priori* probabilities are to be taken into account, we use the following equation to compute the conditional probabilities,

$$P(j/X) = \frac{\pi_j P(X/j)}{\pi_1 P(X/1) + \pi_2 P(X/2) + \ldots + \pi_n P(X/n)}, \tag{8.2}$$

where $\pi_j P(X/j)$ is the joint probability that a randomly selected individual belongs to population j (i.e. that portion of the population suffering from disease j) and displays response pattern X.

The $P(X/j)$ values are calculated in exactly the same way as before. Suppose we again have a response pattern $X = (0\ 1\ 1)$, then as we have seen the values of $P(X/j)$ are,

$P(X/1) = 0.045,\quad P(X/2) = 0.009,\quad P(X/3) = 0.019,\quad P(X/4) = 0.015.$

Application of equation (8.2) gives the following result,

$$P(1/X) = \frac{\pi_1 P(X/1)}{\pi_1 P(X/1) + \pi_2 P(X/2) + \pi_3 P(X/3) + \pi_4 P(X/4)}$$

$$= \frac{0.16 \times 0.045}{0.16 \times 0.045 + 0.30 \times 0.009 + 0.23 \times 0.019 + 0.31 \times 0.015}$$

$$= 0.38$$

Similarly, $P(2/X) = 0.14$, $P(3/X) = 0.23$ and $P(4/X) = 0.25$.

Thus, individuals displaying a response pattern $X = (0\ 1\ 1)$ are most likely to belong to class 1 (mental and nervous diseases). This is, of course, the same conclusion we reached in the equal base rates model. For this particular pattern, therefore, the inclusion of the actual base rates in the analysis has not altered the classification decision. However, the probability of making the correct decision has decreased from 0.51 when equal base rates were assumed to 0.38 using the unequal base rates model. But in the absence of any knowledge about sex, age and type of employment, the probability of correctly classifying a randomly selected individual as belonging to the class 1 disease category is only 0.16 (the value of π_1, corresponding to the actual relative proportion). When information on the three variables is taken into account, however, the probability of correct classification is more than doubled, to 0.38.

The complete set of classification decisions and their associated probabilities of being correct are given in Table 8.10.

TABLE 8.10 *Variable response pattern and corresponding chronic sickness classification, together with probability of correct assignment (unequal base rates model)*

Response pattern	111	110	101	100	011	010	011	000
Disease classification	3	3	2	2	1	1	4	4
Probability	0.33	0.36	0.34	0.32	0.38	0.42	0.35	0.33

Some points of interest arise out of the results shown in Table 8.10. Firstly, the disease classification decisions are completely different in the case of three of the four patterns (those corresponding to (1 1 1), (1 1 0) and (1 0 0)), as a consequence of taking into account the *a priori* probabilities. Moreover, although the remaining sickness classifications are the same (the patterns (1 0 1), (0 1 1), (0 1 0), (0 0 1) and (0 0 0)) the probabilities of correct classification are somewhat different. Secondly, as we can see from Table 8.10, it appears that all the correct classification decisions depend on the responses to the first two variables. The type of employment variable appears to be irrelevant, whereas, with the equal base rates model, the employment variable was a factor distinguishing groups displaying patterns 101 and 100, i.e. males over 44.

The unequal base rates model clearly differentiates between each of the four sickness categories. As Table 8.10 shows, females aged 44 or less will most probably suffer from mental and nervous diseases, but males aged 44 or less are most likely to experience respiratory diseases. On the other hand, males aged over 44 are likely to suffer from heart and circulatory diseases, whilst females of the same age group from musculo-skeletal illnesses. Employment, as we have indicated, plays no part in the classification process.

This example has illustrated the use of the pattern probability model and enables us to decide which chronic disease, out of four possible such illnesses, is most probable should an individual become chronically sick, in the light of information on sex, age and type of employment.

However, the categories chosen are neither mutually exclusive nor exhaustive since there are other categories of chronic illness which occur (such as digestive disorders) and, furthermore, individuals may suffer more than one disease. In the present example, a non-exhaustive set of disease categories was used for the purpose of keeping the analysis simple. It would have been possible to have constructed the problem such that the categories formed an exhaustive set* (there are 13 such illnesses recorded in the General Household Survey). Of course, if we are simply interested in discriminating between the illness categories we have chosen (or any other subset) then this problem does not arise. It did arise in the chapter on discriminant analysis, where the use of that method enabled us to say to which of several categories an individual was most likely to belong, but where in actual fact it was possible that an individual did not belong to any. In the example used in the discriminant analysis section, an individual was classified on the basis of certain information as being either Democrat or Republican, when in fact he or she might belong to the Socialist Party, or to no party at all.

8.9 Latent structure analysis

Latent structure analysis is similar to factor analysis in that the relationships between variables are explained in terms of *underlying* variables, except that the method is used specifically with problems having qualitative measures on variables and no use is made of correlation coefficients. The method divides the sample into *latent* classes, none of which exhibits the associations evident in the total sample. It is similar in some ways to other techniques which aim to discriminate between groups, such as the pattern probability model discussed above and the method of discriminant analysis. The essential and important difference is that the model is used for discrimination between *latent* rather than manifest classes. We illustrate the technique with a simple example which involves only two latent classes.

8.10 An example with two latent classes and three dichotomous items

We take our fictitious example from a study of the attitude of school-leavers to various possible occupations. A sample of 216 such children are asked whether they would like to enter any of three professions: medicine, engineering and teaching, designated M, E and T respectively. Suppose the pattern of responses is as shown in Table 8.11, where a 1 indicated a positive 'yes' response and 0 a negative response. For example, the contents of row three of the table mean that 30 children indicated that they would like both medicine and teaching but none of these responded favourably to an engineering career.
Taking each pair of professions in turn, we can draw up the following three contingency tables, shown in Tables 8.12 (*a*), (*b*) and (*c*). In these tables, M_1 indicates that the child would like to take up medicine as a profession, M_0 that he or she would not; similarly for E_1 and E_0, and T_1 and T_0. The entries in these tables are expressed in terms of proportions rather than frequencies.

---

*In a sense we might consider the data in the General Household Survey to form a mutually exclusive set, since respondents were asked which was the most seriously restricting complaint from which they suffered.

TABLE 8.11 *Career preferences of 216 school-leavers*

Medicine, M	Engineering, E	Teaching, T	Number
1	1	1	65
1	1	0	31
1	0	1	30
1	0	0	45
0	1	1	5
0	1	0	7
0	0	1	11
0	0	0	22
		Total	216

TABLE 8.12 *Pair-wise contingency tables for career preferences*

	M_1	M_0		M_1	M_0		E_1	E_0
E_1	0.44	0.06	T_1	0.44	0.07	T_1	0.32	0.19
E_0	0.35	0.15	T_0	0.35	0.13	T_0	0.18	0.31
	0.79	0.21		0.79	0.20		0.50	0.50
	(*a*)			(*b*)			(*c*)	

In Table 8.12(*a*), we see that 44 per cent of the sample of children would like both medicine and engineering as careers, 6 per cent would like engineering but not medicine, 15 per cent would prefer not to enter either career, and so on.

An examination of the responses shown in these contingency tables reveals that associations exist between the choices of the children for different career combinations. For example, from 8.12(*a*), 0.44/0.79 = 56 per cent of children who would like medicine would also like engineering, but only 0.06/0.21 = 27 per cent of those who would *not* like medicine would like engineering. Clearly an association exists between preferences for these two professions amongst the group of school-leavers as a whole. From 8.12(*b*), 56 per cent of children who would like medicine would also like teaching, whilst only 35 per cent of those who would not like medicine would like teaching. From 8.12(*c*), 64 per cent of children would like both engineering and teaching, whilst only 38 per cent of those who would not like engineering would like teaching.

Can the association between these observed categories be explained in terms of some underlying or 'latent' variable of variables? The purpose of a latent structure analysis is to account for or explain the observed associations between children's choices in terms of some underlying unobserved variable. In one sense, then, it has an analogous function to factor analysis, which also seeks to explain observed associations (correlations) in terms of underlying unobserved variables. The nature of this latent variable is such as to divide the sample of children into categories in such a way that the observed associations disappear. The observed associations can come about because they actually do represent associations in each individual in the sample (i.e. in each child's mind there may be an association of preferences for the various professions) or because the sample is made up of several quite distinct groups. Latent structure analysis will tell us whether the latter is the case, i.e. whether the associations are more apparent than real.

In our example, it will prove sufficient to divide the children into two groups, G_1 and G_2. Contingency tables similar to those in Table 8.13 can be drawn up for each group separately, and these are shown in Tables 8.13 (*a*) – (*f*). The way in which these tables are arrived at, i.e. the method of latent structure analysis, will be described in detail a little later; at this stage we want rather to discuss in a general way the sort of results the analysis produces and how we can use it for classification purposes.

TABLE 8.13 *Contingency tables of career preferences of two groups of school-leavers*

	Group G_1				Group G_2		
		M_1	M_0		M_1	M_0	
(*a*)	E_1	0.74	0.04	E_1	0.12	0.08	(*d*)
	E_0	0.21	0.01	E_0	0.50	0.30	
		0.95	0.05		0.62	0.38	
		M_1	M_0		M_1	M_0	
	T_1	0.69	0.04	T_1	0.17	0.11	
(*b*)	T_0	0.26	0.01	T_0	0.45	0.27	(*e*)
		0.95	0.05		0.62	0.38	
		E_1	E_0		E_1	E_0	
	T_1	0.57	0.16	T_1	0.96	0.92	
(*c*)	T_0	0.21	0.06	T_0	0.14	0.58	(*f*)
		0.78	0.22		0.20	0.80	

For example, Table 8.13 (*a*) shows that 78 per cent (0.74/0.95) of the children in group 1 who express a favourable response to medicine also react favourably to engineering, whilst approximately the same proportion, 80 per cent (0.04/0.05)., who are *not* attracted to medicine, are attracted to engineering. Thus, a favourable attitude towards an engineering career is independent of a liking for a medical career. The other tables confirm this independence.

The distinction between the two groups is perhaps more obvious if we examine the entry in the lower right-hand corner of each table; for example, in

Table 8.13 (*a*) only 1 per cent of this group do not like either engineering or medicine, whereas in Table 8.13 (*d*) the corresponding figure for the G_2 group is 30 per cent. Similar differences can be observed for other cells in the other tables.

Thus, the originally observed associations can be put down to the fact that the G_1 children tend to like all the professions under examination, whilst the G_2 children do not. For example, in Table 8.13 (*a*), we see that 74 per cent of the G_1 group are attracted by *both* medicine and engineering but only 1 per cent like neither, whilst from 8.13 (*d*) we see that only 12 per cent of the G_2 children are attracted by both professions and 30 per cent would like neither. The previously noted associations between a liking for one career and a liking for the other has disappeared now that the children have been separated into two groups. When the sample of children is considered as a whole, the effect is to produce an apparent association between a liking for the two careers of medicine and engineering; similar considerations apply to the other pair-wise career combinations. These apparent associations do not apply to individuals but results from the amalgamation of two quite properly distinct groups.

This analysis prompts the question – what is the distinction between the two classes G_1 and G_2? Only further investigation can answer this. It may well be found that the G_1 children are those whose academic careers have been more successful, for instance.

We presented the results of the latent structure analysis in Table 8.13 above without any indication of how they were arrived at. We now want to discuss the method in more detail.

8.11 The basic model

In this section we present briefly the theory underlying the basic latent structure analysis model: in the following section we will describe how the parameters of this model are obtained and, finally, we will discuss the use and interpretation of the results. Let us suppose we have a situation where there are three dichotomous items, A, B and C, and two underlying or latent classes, G_1 and G_2. The model can be represented by the following set of equations, usually referred to as the accounting equations

Firstly, $P(A_1) = P(G_1)\,P(A_1/G_1) + P(G_2) \quad (A_1/G_2)$,

where $P(A_1)$ is the probability of giving a positive response to item A (e.g. a 'yes' or '1' to medicine in our previous example); $P(G_1)$ is the probability of belonging to class 1; and $P(A_1/G_1)$ is the probability of giving a positive response to A given that the respondent belongs to class 1 (this equation follows from the basic tenets of conditional probability theory).

Similar equations can be established for $P(B_1)$ and $P(C_1)$ and we can express these in a more general form, thus,

$$P(N_1) = \sum_j P(G_j) \times P(N_1/G_j), \tag{8.5}$$

where $P(N_1)$ is the probability of a positive response to item N.

Secondly, the probability of a positive response to both items A and B is given by,

$$P(A_1 \text{ and } B_1) = P(G_1)P(A_1/G_1) \times P(B_1/A_1 \text{ and } G_1) + P(G_2) \times P(A_1/G_2) \times P(A_1 \text{ and } G_2).$$

But since we are interested in dividing the sample into latent classes in such a way that responses to items become independent, as illustrated in the contingency Tables 8.13, then this equation (using the probability rule for independent events) reduces to,

$$P(A_1 \text{ and } B_1) = P(G_1) \times P(A_1/G_1) \times P(B_1/G_1) + P(G_2) \times P(A_1/G_2) \times P(B_1/G_2) \tag{8.6}$$

Similar expressions can be found for $P(A_1 \text{ and } C_1)$ and $P(B_1 \text{ and } C_1)$. In general, we can express these probability relationships with the equation,

$$P(N_1 \text{ and } M_1) = \sum_j P(G_j) \times P(N_1/G_j) \times P(M_1/G_j). \tag{8.7}$$

Finally, the joint probability of a positive response to all three items, once again assuming independent responses to each item, is given by,

$$P(A_1 \text{ and } B_1 \text{ and } C_1) = \sum_j P(G_j) \times P(A_1/G_j) \times P(B_1/G_j) \times P(C_1/G_j), \tag{8.8}$$

which can also be expressed in the more general form

$$P(N_1 \text{ and } M_1 \text{ and } L_1) = \sum_j P(G_j) \times P(N_1/G_j) \times P(M_1/G_j) \times P(L_1/G_j) \tag{8.9}$$

The accounting equations (8.6) to (8.9) constitute the basic structure for the two latent-classes three-item model. If we insert the observed proportions of positive responses on to the left-hand side of equations (8.6) to (8.9) we can solve them to obtain values for the terms on the right-hand side. This we will now do.

8.12 Solution of accounting equations

In this treatment of the latent structure model we will not attempt to explain the theory underlying the solution of the accounting equations; the interested reader is referred to an eminently readable account in Lazarfeld and Henry [1968]

The steps in the estimation of the latent structure parameters, i.e. $P(G_1)$, $P(A_1/G_1)$ etc., are as follows:

(1) Determine from the observed values in Table 8.11 the values of the proportions $P(A_1)$, $P(B_1)$, $P(C_1)$, $P(A_1B_1)$, $P(A_1C_1)$, $P(B_1C_1)$ and $P(A_1B_1C_1)$.

(2) Solve the quadratic equation:

$$t^2 - bt + c = 0, \tag{8.10}$$

where

$$b = [(D - E)/F + 1]$$

$$c = D/F$$

$$D = P(A_1B_1C_1) \cdot P(C_1) - P(A_1C_1)P(B_1C_1)$$

$$E = [1 - P(C_1)]\,[P(A_1B_1) - P(A_1B_1C_1)] - [P(A_1) - P(A_1C_1)]\,[P(B_1) - P(B_1C_1)]$$

and

$$F = P(A_1B_1) - P(A_1)P(B_1)$$

The two values for t in the solution of equation (8.10) give us $P(C_1/G_1)$ and $P(C_1/G_2)$ and once these two values are known, the remaining parameters can be obtained from the following equations:

$$P(G_1) = [P(C_1) - P(C_1/G_2)]/[P(C_1/G_1) - P(C_1/G_2)] \tag{8.11(a)}$$

$$P(G_2) = [P(C_1/G_1) - P(C_1)]/[P(C_1/G_1) - P(C_1/G_2)] \tag{8.11(b)}$$

$$P(A/G_1) = [P(A_1C_1) - P(A_1)P(C_1/G_2)]/[P(C_1) - P(C_1/G_2)] \tag{8.12(a)}$$

$$P(B_1/G_1) = [P(B_1C_1) - P(B_1)P(C_1/G_2)]/[P(C_1) - P(C_1/G_2)] \tag{8.12(b)}$$

$$P(A_1/G_2) = [P(A_1C_1) - P(C_1/G_1)]/[P(C_1) - P(C_1/G_1)] \tag{8.13(a)}$$

$$P(B_1/G_2) = [P(B_1C_1) - P(B_1)P(C_1/G_1)]/[P(C_1/G_1)] \tag{8.13(b)}$$

The above equations can only be solved when $P(A_1B_1) - P(A_1)P(B_1) \neq 0$ i.e. when associations exist between responses on A and B, and similarly when $P(A_1C_1) - P(A_1)P(C_1) \neq 0$ and when $P(B_1C_1) - P(B_1)P(C_1) \neq 0$. Clearly, if all of these expressions were zero there would not be any need for a latent structure analysis since there would be no associations between the responses to explain.

In our example, the solution of equation (8.10) proceeds as follows:

(1) The required proportions $P(A_1)$, $P(A_1B_1)$ etc., calculated from the observations of Table 8.11, are shown in tabular form in Table 8.14.

For example, $P(A_1) = (65 + 31 + 30 + 45)/216 = 0.792$

$P(B_1C_1) = (65 + 5)/216 = 0.324$

$P(A_1B_1C_1) = 65/216 = 0.301$

TABLE 8.14 *Calculation of probabilities for accounting equations*

	A_1	B_1	C_1
A_1	0.792	0.444	0.440
B_1		0.500	0.324
C_1			0.514
and	$P(A_1B_1C_1) = 0.301$		

(2)

$$D = [(0.301) \times (0.514) - (0.440) \times (0.324)] = 0.012$$
$$E = (1 - 0.514)(0.444 - 0.301) - (0.792 - 0.440)(0.500 - 0.324) = 0.008$$
$$F = 0.444 - 0.792 \times 0.500 = 0.048.$$

Thus, $b = [(0.012 - 0.008)/0.048] + 1 = 1.083,$

and $c = D/F = 0.012/0.048 = 0.250;$

thus, $t = 0.749 \text{ or } 0.334$

As noted above, the two solution values for t gives us values for $P(C_1/G_1)$ and $P(C_1/G_2)$. It is not important which value we take for these; let us arbitrarily assign $P(C_1/G_1) = 0.749$ and $P(C_1/G_2) = 0.334$.

We can now apply the system of equations (8.11) to (8.13) as follows.

From (8.11(*a*)), and from (8.11(*b*))

$$P(G_1) = (0.514) - (0.334)/(0.749 - 0.334) = 0.434,$$
$$P(G_2) = (0.749) - (0.514)/(0.749 - 0.334) = 0.566$$

Of course, $P(G_1) + P(G_2) = 1.000$, so only one of these two needs to be calculated.

From (8.12(*a*)) and (8.12(*b*)),

$$P(A_1/G_1) = 0.975 \text{ and } P(B_1/G_1) = 0.872,$$

and from (8.13(*a*)) and (8.13(*b*)),

$$P(A_1/G_2) = 0.652 \text{ and } P(B_1/G_2) = 0.215$$

These calculated parameters are summarised in Table 8.15.

TABLE 8.15 *Calculated probabilities*

Class	$P(G_j)$	$P(A_1/G_j)$	$P(B_1/G_j)$	$P(C_1/G_j)$
1	0.434	0.975	0.872	0.749
2	0.566	0.652	0.215	0.334

These values can be used to reproduce the original observed proportions in Table 8.12 by the use of equations (8.5) to (8.9). There is nothing in the algebra of the method to prevent the probabilities assuming 'impossible' values, i.e. greater than 1.000 or negative; if such a result should occur it can only mean that the original hypothesis of two latent classes is untenable and the researcher would start the analysis afresh assuming perhaps three latent classes.

It is clear from Table 8.15 that the most likely response of individuals in class 1 (those with perhaps the longest education) will be (1 1 1), i.e. a positive response on all three items, whilst those individuals who belong to group 2 are most likely only to like medicine and respond negatively to engineering and teaching, producing a response pattern of (1 0 0).

8.13 Classification of respondents

One of the more obvious applications of latent structure analysis is the most likely classification of individuals on the basis of their responses to the three items, and knowledge of the parameters arising from the latent structure analysis, i.e. such as those contained in Table 8.15. For example, suppose one school-child responds yes, yes, no, i.e. (1, 1, 0), to the three professions, then the proportion of the sample who respond in such a way, we can tell from the original sample, is 0.444 (from Table 8.14). From Table 8.15, the probability that a person belongs to class 1 and gives this response is $P(G_1) \times P(A_1/G_1)\, P(B_1/G_1) = 0.369$, and the probability that he belongs to class 2 and gives this response is $P(G_2) \times P(A_1/G_2) \times P(B_1/G_2) = 0.079$. Since $P(A_1B_1) = 0.369 + 0.079 = 0.448$, which is the same as 0.444 (allowing for rounding error), it follows that 36.9 per cent of the respondents produce that pattern and belong to class 1, whilst only 7.90 per cent belong to class 2, and hence we would classify such a respondent as a member of class 1. The probability of misclassification of such a respondent is 0.079/0.448 = 0.176 or 17.6 per cent. In the foregoing account, we have chosen an example involving three dichotomous items (*A, B* and *C* could only take the values 1 or 0), and in which a solution involving only two latent classes could be obtained. These restrictions are not inherent in latent structure analysis, however: the method can be applied to problems with more than three items (which in turn can be polyotomous) and in which the analysis can involve more complex solutions than simply two latent classes.

The interested reader is referred to the book by Lazarfeld and Henry [1968]. However, these issues are beyond the scope of this introductory text.

8.14 An example from the literature

This practical example of the technique of latent structure analysis is due to Miller and Butler [1966] and is taken from the field of sociology. Miller and Butler were concerned with the suitability of Srole's interview schedule for the measurement of anomie. In particular, they were concerned to test the scaling properties and typological construction of Srole's five items.

Samples from two different projects were available for analysis. The first was a survey of 981 households in metropolitan Los Angeles and the other a similar sample (231 households) of a suburban city: questions relating to anomie were administered in both cases as part of a larger investigation. Dichotomised responses on the items were available, a 'positive' response indicating an inclination towards anomie. The responses in both samples were then factor analysed with the result that one factor was sufficient to account for most of the variation in each case, i.e. the five anomie items can all be explained by this single factor. In other words, the responses appear to be unidimensional, which suggests a continuous anomie scale going from completely anomic to completely eunomic. However, a latent structure analysis revealed that there were *two* distinct latent classes which, the authors suggest, could be designated as anomia and eunomia; the first class tends to answer all items positively, the second rarely does so. The different proportions between the two samples, 45 per cent anomic in metropolitan Los Angeles, 36 per cent in the suburban city, reflect anticipated differences between the two types of area.

As a test of the model, misclassification of only 7 per cent indicated that the model fitted both samples well.

The authors conclude that their analysis indicates two polar types, anomic and eunomic, which results in a simplification of the anomia scale. This could reduce the scale to a very few items, rather than extending it as Srole has suggested.

9 Concluding Remarks and Overview

9.1 Aims of factor analysis

The main aims of factor analysis can be re-iterated here:

(*a*) To produce a more parsimonious description within a domain of study. For example, in the body-size problem (Chapter 2), nine variables measuring various aspects of body-size were reduced to three, giving approximately the same amount of information.

(*b*) To test theories about the interrelationships between variables. For example:

  (i) in the analysis of intellectual ability (Chapter 4), the theory that out of eight tests of mental ability three factors (convergent, non-verbal divergent and verbal divergent thinking) would emerge could be tested;

  (ii) in the body-size problem, the theory that there are distinct body-types (short and stocky versus tall and lean) could be tested.

(*c*) To establish functional relations between variables (one of the goals of science). Factor analysis can be employed to isolate variables which it may not be possible to measure directly, but which can be computed from a set of observable and directly measureable but otherwise unsatisfactory measures. This involves factor measurement which will be described in Section 9.2.

(*d*) To analyse people or objects into types, which will be described in Section 9.3.

(*e*) Finally, as a preliminary to regression analysis, to analyse the factorial structure of criterion variables, and hence point the way to those variables which are most likely to be usefully included in a regression equation. This will be discussed below in Section 9.4.

Following discussion of these points, the remainder of the chapter presents a brief description of some multivariate techniques which may be useful to the behavioural scientist but which have not been discussed because of lack of space, and some guidelines are laid down to aid the investigator in his choice of technique.

9.2 Factor measurement

Once an investigator has established that his original set of variables can be reduced to a small set of factors, he may be interested in assigning a score on each factor to each of the individuals or objects which he is studying. There are several ways of doing this, ranging from the very crude and approximate method of using the score which an individual obtains on that variable which best represents a factor (i.e. has the highest loading on that factor) to sophisticated estimation methods based upon the factor loadings of all the variables on a factor. We will content ourselves with presenting two methods which lie some way between the two extremes, have the advantage of being simple to use, and are satisfactory for most purposes. As an illustration, let us take the orthogonal rotation of the body-size problem tabulated in Table 2.2.

Firstly, each variable is scaled to the same mean and standard deviation (say to standard scores, i.e. mean of 0 and a standard deviation of 1). Suppose we have two individuals, X and Y, who obtain standard scores on the nine variables involved in the body-size problem. A score on each of three factors can be obtained for each individual by application of the formula below (9.1),

$$F_{jx} = \sum_i z_i \times f_{ij}, \qquad (9.1)$$

where F_{jx} is the factor score of individual X on factor j, z_i is the standard score of the individual X on variable i and f_{ij} is the factor loading of variable i on factor j. This formula gives due weight to those variables which have high loadings on the factor and minimises the effect of variables with low loadings. Scores are converted to standard scores because in raw score form, variables with large standard deviations would have an inordinately large effect on the factor measure. Factor scores for each individual obtained by this method are entered in Table 9.1. The advantage of this method is that all variables and factor loadings are taken into account. One possible disadvantage is that the investigator may be misled into overemphasising the importance of differences in factor loadings since, after all, the size of these is dependent on rotation, and we have seen in Chapter 2 that this is to some extent arbitrary in a relatively ill-defined field of study.
In Chapter 2, we identified F_1 as 'body-bulk', F_2 as 'leg length' and F_3 as 'trunk length'. Thus individual X is bulkier but has shorter dimensions both above and below the waist than individual Y, although most of the difference in height is attributable to the vast difference in 'trunk length'.

As was stated in Section 9.1, one of the goals of science is to establish functional relations between variables. In the behavioural sciences, many variables are factorially complex, which implies that when they are used in experimental designs or regression analyses, we are in fact compounding variables that appear to be unitary but are in fact made up of several components. Factor analysis can help in the search for variables which are factorially pure, i.e. chiefly measures of one thing, or factor measures themselves can be used in investigations instead of the original variables. Incidentally, the employment of factor measures in subse-

TABLE 9.1 *Factor scores for each individual*

Person	Variable								
	1	2	3	4	5	6	7	8	9
X	1.0	0.5	0.4	0.6	1.2	1.5	2.0	0.0	1.0
Y	2.0	2.1	1.8	0.9	1.0	0.0	0.5	0.1	0.0

Person	Factor score		
	F_1	F_2	F_3
X	3.13	2.87	2.03
Y	1.78	3.24	5.10

quent investigations can help enormously with the rotation problem. In Chapter 3, where the rotation problem was discussed, we mentioned that, from the mathematical point of view, or if we are only interested in parsimony, one rotation is as good as any other. But if we are interested in locating variables which are relatively pure measures of a factor, or in devising ways of combining impure variables in such a way as to get the best possible measure of a factor, then the position of rotated axes is very important. From the scientific investigator's point of view then, some rotations will be useful for extending knowledge in a field of study, and some will be useless. If investigations involving factor measures as variables produce reliable functional relationships with other variables then we can be confident that the position of rotated axes is useful for scientific purposes, even if the tenets of simple structure are violated. Moreover, if this is the case, the factor is essentially a construct that aids in the organisation of scientific knowledge in a field. The diagram below (Fig. 9.1) shows how this organisation is

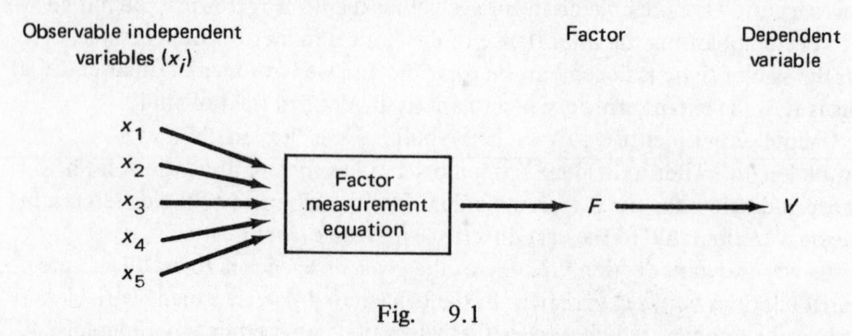

Fig. 9.1

achieved. Without the organising construct or factor, we have five functional relations, generalisations or laws, each of the form $V = f(x_i)$, some or all of which may be unsatisfactory because each x_i measures several things besides F. With the organising construct or factor and the knowledge of how each x_i is related to it, we can replace the five original generalisations by one, $V = f(F)$, which is not only more parsimonious but apt to be more reliable if F is less complex, i.e. a unitary variable.

In this connection, it may be noted that it might be advantageous to employ orthogonal rotation if factors are to be used in subsequent investigations, even if oblique rotations are necessary to achieve simple structure. This is because oblique factors are correlated, and hence if one such factor is used in a subsequent investigation, its effects are compounded with any correlated factors, making interpretation difficult.

9.3 *Q*-type factor analysis

The type of factor analysis with which this book has thus far been concerned is known as R-type analysis which is concerned with analysing the relationships between variables and discovering groups of similar ones; each variable is a measure obtained from a large sample of individuals or objects. In Q-type factor analysis, it is the individuals who are factor-analysed. Each individual is given a score on each variable and these scores are then correlated, i.e. the scores for each individual are then correlated with those of each other individual. Just as in R-type factor analysis, large samples of individuals are required to produce stable correlations, so in Q-type factor analysis a large number of variables are required to achieve the same result. Each variable in a Q-type analysis should be scaled to standard-score form first, otherwise wide discrepancies between the relative magnitudes of different variables will spuriously inflate the correlations between individuals. The purpose of a Q-type factor analysis is to analyse relationships between individuals (or objects) and to locate clusters of similar individuals. Thus it is useful in discovering 'types' of individual or object in a heterogeneous sample.

This type of factor analysis has obvious applications in psychology (to differentiate personality types, for example), medicine (e.g. to differentiate disease entities), and many other areas. In the body-size problem discussed in Chapter 2, it was hypothesised that at least two relatively distinct body-types existed. A Q-type analysis could be applied to this sort of problem in order to test the hypothesis, although ideally many more than nine original variables would be required in order to obtain reliable correlations.

One potentially useful application of Q-technique is in the area of anthropology. An example of this application can be found in Pfeiffer [1972]. A set of measurements could be taken on, say, a collection of skulls; a Q-type analysis might then be used to divide the collection into samples of different types, perhaps of different races or species.

Another application is in the area of repertory grid data. In Chapter 8, an R-type factor analysis was carried out on repertory grid data to identify clusters of similar variables (they were called 'attributes' or 'constructs'). However, the scores on each construct for each of the elements (in the example quoted, the politicians) could have been intercorrelated and a Q-type analysis carried out. This would have enabled the investigator to divide the total set of politicians into several different types, each possessing a similar set of attributes.

9.4 Factor analysis and multiple regression

There exists a certain similarity between factor analysis and multiple regression analysis, as the equations below indicate. The object of multiple regression analysis is to find an equation of the following form:

$$y = \beta_1 X_1 + \beta_2 X_2 \ldots + \beta_i X_i \ldots + \beta_n X_n, \tag{9.2}$$

where y is the 'criterion' variable in standard-score form, the X_i's are 'predictor' variables in standard-score form, and the β_i's are regression coefficients, or weights, whose sizes determine the relative importance of the predictor variables. Once these weights are known, the values that an individual obtains on each predictor variable can be used to predict his score on the criterion variable (e.g. academic success, job satisfaction, accident proneness, etc.).

The basic equation in factor analysis is:

$$z_{jk} = a_{1j}F_{1k} + a_{2j}F_{2k} + \ldots + a_{ij}F_{ik} + \ldots + a_{sj}S_{sk}, \tag{9.3}$$

where z_{jk} is the standard score of person k on variable j, a_{ij} is the factor loading (or correlation) of variable j on factor i, F_{ik} is the factor score of person k on factor i in standard-score form, a_{sj} is the factor loading of variable j on the specific factor and S_{sk} is the factor score of person k on the specific factor. The specific factor (which includes also an error component) represents what the variable z measures, that it does not measure in common with any of the other variables. If the communalities in the analysis are high, then the proportion of variance in each variable attributable to specific factors may be negligible. In a principal component analysis, all of the variance is treated as though it were common variance and the last term in equation (9.3) disappears. Thus, we can predict a person's score on any variable (analogous to the criterion variable in multiple regression) from a knowledge of the factor loadings (analogous to the weights in multiple regression) and the person's scores on the common factors (analogous to predictor variables). In practice, most of the variance is accounted for by the first few common factors, and the less important factors can often be ignored.

One basic difference between equations (9.2) and (9.3) is that the predictor variables in (9.2) are all directly observable and measurable, whereas the factors in (9.3) are all derived measures that are not directly observable in themselves, but have to be estimated from their component observable variables. Thus it would be difficult to employ a factor equation to predict a person's score on variable j, since a knowledge of his score on this variable may well be necessary to obtain his factor scores. The aims of the two models are somewhat different. The basic difference in the hypothesis made by the two methods are as follows. Factor analysis assumes interdependence of variables whilst multiple regression analysis assumes one variable (the criterion) dependent on the others.

As stated in Section 9.1, factor analysis can be a useful preliminary to a multiple regression analysis. One problem with the latter type of investigation is knowing which variables, from what might be a large set, to choose, which will turn out to be valid predictors of the criterion variable. Including the latter in a factor analytic study, however, should throw light on its factor structure, and hence those variables which load highly on the relevant factors should prove useful predictor variables. Moreover, with knowledge of the factor structure of the criterion, it may be possible to find a better criterion that is a more valid measure for the purposes of the investigation.

9.5 An overview

The aim of this final section is to aid the investigator in his choice of the techniques which have been described in this book, and to show the interrelations between these techniques. We will begin with a set of questions which the investigator may wish to answer.

Questions

1. Do we wish to reduce the number of variables in a domain of study, or isolate the important ones, or examine their interrelationships, or produce a more parsimonious description of a sample of individuals or objects?
2. Do we wish to construct a theory about the nature of a construct such as intelligence, personality, criminal behaviour, conflict, etc., but are not sure about its component parts?
3. Do we wish to test a definite hypothesis within a domain regarding its factorial composition, e.g. can mental ability be usefully regarded as being made up of three (or any other number) relatively distinct components?
4. Do we wish to test a hypothesis about the functional relationship between two (or more) variables (e.g. if we increase A, does B increase also), but are aware that the measures we have available are not satisfactory measures of A and/or B?
5. Do we wish to carry out a multiple regression analysis, i.e. to predict scores on one variable from those on a set of others, but are not sure which predictors to employ and/or which criterion variable to employ?
6. Do we wish to investigate the possibility that a population of individuals or objects is composed of several 'types' with distinct characteristics?
7. Do we wish to locate clusters of similar or related variables?
8. Do we wish to find an efficient way of discriminating between several groups on the basis of several variables?
9. Do we wish to find an efficient way of classifying individuals into one of several groups on the basis of several variables?
10. Do we wish to analyse relationships between variables measuring aspects of objects or individuals, but are uncertain which variables are important to the sample we are studying and/or uncertain about the objects or individuals that will be measured by these variables? This kind of problem arises when we wish to analyse judgements people make about objects but are unclear what attributes people use in making these judgements.

Appropriate techniques

1 and 2. Factor analysis or principal component analysis; the latter only where there are a large number of variables or when we can expect communalities to be high. The centroid or square-root method only when computing facilities are unavailable, otherwise the principal factor analysis method is recommended. These methods assume that the variables in which the investigator is interested are known and can be measured, preferably quantitatively. In the special case where one wishes to account for the associations between variables in contingency tables, latent structure analysis might be employed.

3. If we have definite *a priori* assumptions about the factorial composition of a set of variables, or which of them form clusters of similar variables, the multiple groups method is an appropriate technique.

4. One of the methods of factor analysis described earlier is an appropriate starting point in order either to identify factorially pure variables, or to use factors rather than original variables to specify the constructs whose functional relationships are to be examined.

5. As described earlier, one of the methods of factor analysis can be employed as a starting-point.

6. For this sort of problem, a *Q*-type factor analysis can help when there are a sufficient number of quantitative variables. With contingency table data, latent structure analysis can be an appropriate technique.

7. Any method of factor analysis, with appropriate subsequent rotation, can be used for the location of clusters. Elementary linkage analysis can also be employed for simplicity.

8. If one wishes to merely discriminate between several groups and the number of discriminant functions involved is immaterial (i.e. parsimony is unimportant), then discriminant function analysis is appropriate. However, if parsimony of the decision procedure is required, then canonical discriminant analysis is to be preferred. Both methods require quantitative measurement of the relevant variables. With qualitative data, the pattern probability model is appropriate. All these methods assume that the investigator has a clear idea of the groups between which he wishes to discriminate. If this is not the case, then a combination of techniques may be necessary, i.e. *Q*-type factor analysis or latent structure analysis to divide the population into groups or 'types', followed by one of the techniques above in order to develop a decision procedure for discrimination between the groups.

9. The problem of classifying individuals or objects is the converse of the problem of discriminating between groups. Hence the appropriate techniques are those discussed in the previous paragraph.

10. For this kind of problem, if the people in our sample are able to state the kinds of attributes upon which they base their judgements, then the repertory

grid technique followed by factor analysis is an appropriate technique. Alternatively, or if our sample has difficulty in assigning verbal labels to the attributes relevant to their judgements of say, similarity, then a multidimensional scaling procedure followed by factor analysis is appropriate.

9.6 Other multivariate techniques

There are many other multivariate techniques of great potential value to the behavioural scientist, but space does not permit more than a brief mention of some of them here:

(*a*) Multiple regression analysis has been briefly mentioned, and a fuller discussion can be obtained in Kerlinger and Pedhazur [1973].

(*b*) Since so much of the data in the behavioural sciences is cast in the form of contingency tables, techniques of analysing such tables other than latent structure analysis can be found useful and are discussed in Plackett [1974].

(*c*) The reader is probably already familiar with analysis of variance methods involving one dependent variable. When several dependent variables are involved on the same sample, however, multivariate analyses of variance or multivariate analogues of Students' t-test are appropriate. A discussion of such methods may be found in Overall and Klett [1972].

(*d*) In the discussion of Q-type factor analysis, it was pointed out that the analysis revealed similarities between individuals. In actual fact, the similarity between any two individuals stems from the correlation (or similarity in shape) which exists between their score profiles (i.e. they will tend to obtain high scores on one set of variables and low ones on another set). This is irrespective of the difference in absolute magnitude of the differences between their score profiles. Thus a high correlation between score profiles for two individuals is compatbile with a gross dissimilarity between them. Overall and Klett [1972] discuss this problem.

(*e*) As well as R- and Q-types of factor analysis, there are other forms, e.g. the P-technique, which involves correlating measures on several variables on the same individual taken at different periods of time. There are various other lesser-known methods which can be obtained from Catell [1952].

The reader should now have a resonable intuitive grasp of the essentials of several multivariate techniques. If it is desired to go more deeply into the underlying theory and mathematics associated with these techniques then the reader is referred to more advanced texts by Harman [1967], Horst [1965], Lawley and Maxwell [1971], Overall and Klett [1972], Comrey [1973], Tatsuoka [1971] and Van der Geer [1971].

References

B. Ahamad [1967] 'An Analysis of Crimes by the Method of Principal Components', *J. Appl. Stats.*

A. R. Baggeley [1964] *Intermediate Correlation Methods* (New York: Wiley).

D. Bannister and J. M. Mair (eds) [1968] *The Evaluation of Personality Constructs* (Academic Press).

P. M. Blau and O. D. Duncan [1967] *The American Occupational Structure* (New York: Wiley).

J. E. Bledsoe [1973] 'The Prediction of Teacher Competence: A Comparison of Two Multivariate Techniques', *Mult. Behav. Res.*, 8, 3 - 22.

C. Burt and C. H. Banks [1941] 'A Factor Analysis of Body Measurements for British Adult Males', *Ann. Eugen. Lond.*, 13, 338.

R. B. Catell [1967] 'The Theory of Fluid and Crystallised Intelligence Checked at the 5 to 6 year-old Level', *Brit. J. Educ. Psychol.*, 37, 209 - 24.

R. B. Catell [1952] *Factor Analysis* (Harper).

R. B. Catell [1963] 'Theory of Fluid and Crystallised Intelligence: A Critical Experiment', *J. Educ. Psychol.*, 54, 1 - 22.

D. Child [1970] *The Essentials of Factor Analysis* (Holt, Rinehart and Winston).

A. L. Comrey [1973] *A First Course in Factor Analysis* (Academic Press).

C. H. Coombs [1964] *A Theory of Data* (New York: Wiley).

R. Dobson, T. F. Golob and R. L. Gustafson [1974] 'Multidimensional Scaling of Consumer Preferences for a Public Transportation System: An Application of Two Approaches', *Socio-Econ. Plan. Sci.*, 8, 23 - 6.

J. A. Giggs [1973] 'The Distribution of Schizophrenics in Nottingham', *Inst. Brit. Geog. Trans.*, 59, 55 - 73.

W. L. Gramm [1973] 'The Labour Force Decision of Female Married Teachers: A Discriminant Analysis Approach', *Rev. Econ. Stats.*, LV, 341 - 8.

J. P. Guilford [1954] *Psychometric Methods* (McGraw-Hill).

L. Guttman [1954] 'Some Necessary Conditions for Common-factor Analyses', *Psychometrica,* 19, 149 - 61.

H. H. Harman [1967] *Modern Factor Analysis* (University of Chicago Press).

P. G. Hoel [1971] *Elementary Statistics* (New York: Wiley).

P. Horst [1965] *Factor Analysis of Data Matrices* (Holt, Rinehart and Winston).

G. A. Kelly [1955] *The Psychology of Personal Constructs*, Vols I and II (Norton).

F. N. Kerlinger and E. J. Pedhazur [1973] *Multiple Regression in Behavioural Research* (Holt, Rinehart and Winston).

J. B. Kruskal [1964] 'Multidimensional Scaling by Optimising Goodness of Fit to a Non-metric Hypothesis', *Psychometrica*, 29, 1 - 27.

D. N. Lawley [1940] 'The Estimation of Factor Loadings by the Method of Maximum Likelihood', *Proc. R. Soc. Edin.*, Series A, 60, 64 - 82.

D. N. Lawley and A. E. Maxwell [1971] *Factor Analysis as a Statistical Method* (Butterworth).

P. F. Lazarfeld and N. W. Henry [1968] *Latent Structure Analysis* (Houghton Mifflin).

L. L. McQuitty [1957] *Educ. Psych. Meas.*, 17, 207.

C. R. Miller and E. G. Butler [1966] 'Anomia and Eunomia: A Methodological Evaluation of Srole's Anomie Scale', *Amer. Soc. Rev.*, 31, 400 - 6.

J. E. Overall and J. Klett [1972] *Applied Multivariate Analysis* (McGraw-Hill).

J. E. Pfeifer [1972] *The Emergence of Man* (Harper and Row).

A. E. Philip and J. W. McCulloch [1966] 'Use of Social Indices in Psychiatric Epidemiology', *Brit. J. Preventive and Soc. Med.*, 20, 122 - 6.

R. L. Plackett [1974] *The Analysis of Categorical Data* (Griffin).

Social Trends [1972] No. 3 (H.M.S.O.).

M. M. Tatsuoka [1971] *Multivariate Analysis* (New York: Wiley).

The General Household Survey [1973] *Introductory Report* (H.M.S.O.).

L. L. Thurstone [1927] 'A Law of Comparative Judgement', *Psychol. Rev.*, 34, 273 - 86.

L. L. Thurstone [1947] *Multiple Factor Analysis* (University of Chicago Press).

W. S. Torgerson [1958] *Theory and Methods of Scaling* (New York: Wiley).

J. P. Van de Geer [1971] *Introduction to Multivariate Analysis for the Social Sciences* (Freeman).

T. Yamane [1964] *Statistics; An Introductory Analysis* (Harper and Row).

M. Ziegler and T. H. Atkinson [1973] 'Information Level and Dimensionality of Liberalism and Conservatism', *Mult. Behav. Res.*, 8, 195 - 212.

Index